Instrumentation and Control

AWWA MANUAL M2

Third Edition

American Water Works Association

MANUAL OF WATER SUPPLY PRACTICES—M2, Third Edition

Instrumentation and Control

Copyright © 2001 American Water Works Association

All rights reserved. No part of this publication may be reproduced or transmitted in any form or by any means, electronic or mechanical, including photocopy, recording, or any information or retrieval system, except in the form of brief excerpts or quotations for review purposes, without the written permission of the publisher.

Library of Congress Cataloging-in-Publication Data

Instrumentation and control.-- 3rd ed.
    p. cm. -- (AWWA manual ; M2)
  New ed. of: Automation and instrumentation. c1983.
  Includes bibliographical references and index.
  ISBN 1-58321-125-X
    1. Waterworks--Automation. I. American Water Works Association. II. Automation and instrumentation. III. Series.

TD491 .A49 no. M2 2001
[TD487]
628.1 s--dc21
[628.1]
                                              2001055311

Printed in the United States of America

American Water Works Association
6666 West Quincy Avenue
Denver, CO 80235

ISBN 978-158321-1-250

Printed on recycled paper

Contents

List of Figures, v

List of Tables, xi

Foreword, xiii

Acknowledgments, xv

Chapter 1 Introduction 1

    The Water Utility System, 1
    How to Use This Manual, 3
    Reference, 4

Chapter 2 Hydraulics and Electricity 5

    Hydraulics, 5
    Electricity, 18
    References, 39

Chapter 3 Motor Controls 41

    Introduction, 41
    Motors, 41
    Variable Speed Motor Control, 49
    Variable Speed Motor Control Systems, 50
    Motor Control Logic, 52

Chapter 4 Flowmeters 67

    Meter Categories, 67
    Meter Coefficient of Discharge, 68
    Venturi Flowmeters, 69
    Modified Venturis, 74
    Orifice Plate Flowmeters, 74
    Magnetic Flowmeters, 76
    Turbine and Propeller Flowmeters, 80
    Sonic Flowmeters, 84
    Vortex Flowmeters, 86
    Averaging Pitot Flowmeters, 89
    Variable Area Flowmeters, 92
    Open Channel Flow, 94
    General Installation Precautions, 98
    Signal Output and Transport, 99
    References, 100

Chapter 5 Pressure, Level, Temperature, and Other Process Measurements 101

    Pressure, Level, and Temperature, 102
    Electric Power and Equipment Status, 110
    Process Analyzers, 112
    General Considerations, 119
    References, 119

Chapter 6 Secondary Instrumentation **121**

    Introduction, 121
    Signal Standardization, 121
    Signal Power and Transmission, 122
    Transmitters, 124
    Controllers, 124
    Recording and Indicating Hardware, 126
    Function Modules, 128
    Converters, 129

Chapter 7 Telemetry **131**

    Analog Telemetry, 133
    Tone Multiplexing, 137
    Amplitude Modulation Tone, 137
    Frequency Shift Keying Tone, 138
    Communication Media and Channels, 138
    Reference, 142

Chapter 8 Final Control Elements **143**

    Valves, 144
    Valve Summary, 153
    Pumping Systems, 154
    Miscellaneous Final Control Elements, 157

Chapter 9 Basics of Automatic Process Control **161**

    Feedforward Control, 162
    Feedback Control, 163
    Feedforward vs. Feedback Control, 164
    Manual vs. Automatic Control, 165
    Automatic Feedforward Control Methods, 166
    Automatic Feedback Control Methods, 168
    References, 178

Chapter 10 Digital Control and Communication Systems **179**

    Digital Control Systems, 180
    Communication Systems, 188
    Applications and Site Planning, 194
    Technology Trends, 196
    References, 197

Chapter 11 Instrument Diagrams **199**

Glossary, 207

Index, 215

List of AWWA Manuals, 225

Figures

2-1   Pressure in a tank, 7

2-2   Pressure in containers of various shapes, 7

2-3   Water level in an unpressurized system, 8

2-4   Fluid levels in a vacuum system, 8

2-5   Flow velocity as a function of cross-sectional area, 9

2-6   Flow–velocity profiles, 9

2-7   Determination of static pressure, 9

2-8   Water in pipe with pressure, no flow, 10

2-9   Total head, 11

2-10 Elevation head, 11

2-11 Flowing without friction, 12

2-12 Velocity head, 12

2-13 Flowing with friction, 13

2-14 Flowing with friction, 15

2-15 Mechanical leverage compared to hydraulic force, 16

2-16 Hydraulic force, 17

2-17 Differential areas, 17

2-18 Transformer symbol, 24

2-19a Delivery voltage at 480 VAC using electric utility's transformer, 25

2-19b Delivery voltage at 21,000 VAC using water utility's transformer, 25

2-20 Main substation with switchgear, 26

2-21 Complete one-line with load center and motors, 28

3-1   Induction motor rotors, 43

3-2   Motor starter contactor coil, 53

3-3   Motor starter circuit with one switch, 53

3-4   Motor starter circuit with two switches, 54

3-5   Maintained contact switch symbol, 54

3-6   Momentary contact switch symbols, 54

3-7   Momentary start switch circuit, 55

3-8   Control relay coil symbol, 55

3-9   Control relay contact symbols, 55

3-10 Three-wire motor control circuit, 56

3-11	Three-wire motor control circuit with two control locations, 57	
3-12	Ladder diagram with line numbers, 57	
3-13	Status indicating light symbol, 58	
3-14	Motor circuit with indicating lights, 58	
3-15	Selector switch symbol, 58	
3-16	Motor circuit with local–remote switch, 59	
3-17	Hand-off–auto switch, 59	
3-18	HOA motor circuit, 60	
3-19	Float-operated level switch symbol (closes on rising level), 60	
3-20	Float-operated level switch symbol (opens on rising level), 60	
3-21	Automatic pump control off of a float switch, 61	
3-22	Three-wire control using two level switches, 61	
3-23	Three-wire control using two level switches with lock-out–stop switch, 63	
3-24	Three-wire control using two level switches with lock-out–stop switch and a low-level interlock switch, 64	
3-25	Electrical ladder diagram symbol legend, 65	
4-1	The Venturi tube, 69	
4-2	Venturi meter and flow tube, 70	
4-3	Troubleshooting guide for a differential pressure transducer, 72	
4-4	Orifice plate, 75	
4-5	Magnetic flowmeter, 77	
4-6	Example of a troubleshooting flowchart, 78	
4-7	Propeller and turbine meters, 81	
4-8	Troubleshooting procedures for turbine meter, 82	
4-9	Ultrasonic time-of-flight flowmeter, 84	
4-10	Vortex flowmeter, 87	
4-11	Vortex flowmeter troubleshooting guide, 88	
4-12	Averaging Pitot flowmeter insertion tube, 90	
4-13	Variable area flowmeter, 92	
4-14	Common types of weirs, 95	
4-15	Free flow over a weir, 96	
4-16	Parshall flume, 97	
4-17	Typical flow straighteners, 99	
5-1	Bourdon, bellows, and diaphragm pressure sensors, 103	
5-2	Typical LVDT application, 103	
5-3	Diaphragm seal, 104	

5-4	Variable capacitance pressure sensor, 104	
5-5	Float-type, level-sensing system, 105	
5-6	Stage recorder, 106	
5-7	Bubbler, 106	
5-8	Admittance probe, 107	
5-9	Variable resistance level sensor, 108	
5-10	Ultrasonic level sensor, 108	
5-11	Typical temperature elements, 109	
5-12	Thermowell, 110	
5-13	Motor current sensor, 111	
5-14	Light scatter turbidity, 113	
5-15	Surface scatter, 114	
5-16	pH system, 114	
5-17	Immersion and flow-through pH systems, 115	
5-18	Chlorine membrane probe, 116	
5-19	Amperometric chlorine residual analyzer, 117	
5-20	CO_2 buffering, 117	
5-21	Particle counter, 118	
5-22	Streaming current monitor, 118	
6-1	Typical single compressor system, 124	
6-2	Power supply, 125	
6-3	Basic controller, 125	
6-4	Analog indicator, 126	
6-5	Analog and digital indicator, 126	
6-6	Circular recorder, 127	
6-7	Strip chart recorder, 127	
7-1	Telemetering, 132	
7-2	Typical digital telemetering system, 134	
7-3	Schematic of a typical PDM system, 135	
7-4	Nomenclature of frequencies, 140	
8-1	Components of control, 144	
8-2	Solenoid with cylinder actuator, 145	
8-3	Solenoid with details, 145	
8-4	Single-phase motor, 146	
8-5	Pneumatic positioner cut away, 147	

8-6	Electronic positioner circuitry, 147	
8-7a	Rotary valve requires torque, 148	
8-7b	Linear valve requires thrust, 148	
8-8	Piping configurations, 149	
8-9	Control characteristics, 151	
8-10	Butterfly valve, 151	
8-11	Plug valve, 152	
8-12	Gate valve, 152	
8-13	Globe valve, 152	
8-14	Discharge pressure control via series valve, 156	
8-15	Discharge pressure control via bypass valve, 157	
8-16	Pneumatic conveying system, 158	
8-17	Chemical feed system (liquid), 158	
8-18	Chemical feed system (dry), 159	
8-19	Typical rotary paddle volumetric feeder, 159	
8-20	Screw-type volumetric feeder, 160	
8-21	Gravimetric feeder (belt type), 160	
9-1	Generic control loop, 162	
9-2	Feedforward control of chlorine contact channel, 163	
9-3	Feedback control of chlorine contact channel, 164	
9-4	Compound control of chlorine contact channel, 166	
9-5	Generic feedback control timing graph, 169	
9-6a	On–off control of a reservoir, 170	
9-6b	On–off control timing graph, 170	
9-7a	Gap-action control of a reservoir, 171	
9-7b	Gap-action control timing graph, 171	
9-8	Proportional control input/output relationship, 172	
9-9a	Proportional control of a reservoir, 173	
9-9b	Proportional control timing graph, 173	
9-10a	Integral control of a reservoir, 175	
9-10b	Integral control timing graph, 175	
9-11a	Proportional-plus-derivative control of a reservoir, 177	
9-11b	Proportional-plus-derivative control timing graph, 178	
10-1	Digital control system, 181	
10-2	Operating system, 187	

10-3   Layers of communications, LAN, WAN, 190
10-4   Reference model for open system interconnection, 190
10-5   Networks, 192
11-1   General instrument or function symbols, 201
11-2   Function designations for relays, 202
11-3   Standard instrument line symbols, 203
11-4   Example of PI&D loop description, 205

This page intentionally blank.

Tables

6-1   Comparison of electronic and pneumatic systems, 123

10-1 EIA standards, 191

11-1 ISA instruments, 200

11-2 Summary of special abbreviations, 204

This page intentionally blank.

Foreword

The Distribution Division of American Water Works Association authorized the formation of the Committee on Automation and Instrumentation in 1968 and the division's board of trustees set a scope for the committee as follows:

> Assemble and disseminate information on automatic and remote operation and instrumentation of pumping stations and distribution and supply systems, including valves, storage tanks, and booster pumps.

To accomplish part of this assignment, the committee prepared this manual. This third edition of M2 includes basic information on electrical power distribution and updates the automation and instrumentation technology that has changed rapidly since the second edition in 1983. Computer technology in particular is evolving with great speed and will continue to do so. On the other hand, much of the theory and principles remain the same.

This manual is written primarily for the operator of any water utility, large or small, who would not necessarily have technical background but who would be searching for basic explanations and general information. The manual discusses equipment, terms, and expressions an operator encounters wherever electrical systems, automation, and instrumentation are found in water distribution, treatment, and storage systems.

The size and complexity of the centralized water system facilities vary with the size and complexity of the utility. However, the basic principles of electrical power distribution, automation, and instrumentation apply in each case.

The committee, which is now under the Engineering and Construction Division, has been renamed, Instrumentation and Control, and the M2 manual is now titled *Instrumentation and Control*.

This page intentionally blank.

Acknowledgments

The Automation and Instrumentation Committee of the Engineering and Construction Division of the Technical and Educational Council of the American Water Works Association developed this edition of M2. The committee has been renamed Instrumentation and Control. The following committee members assisted:

D. W. Mair, Chair, Cholla Electrical Consultants, Inc., Phoenix, Ariz.
M. J. Okey, Vice Chair, CH2M Hill, Denver, Colo.
H. D. Gilman, Greeley & Hansen Engineers, Huntingdon Valley, Pa.
D. R. Olson, Professional Services Group, Inc., Marshfield, Mass.
E. W. Von Sacken, Colorado Springs Utilities, Colo.
R. K. Weir, Fluids Engineering, Denver, Colo.
E. F. Baltutis, Keystone Controls, Houston, Texas
R. V. Frykman, CDM, Chicago, Ill.
Clarence Hilbrick Jr., Portland Water Bureau, Portland, Ore.
John McDaniel
E. F. Morey, retired

This page intentionally blank.

AWWA MANUAL M2

Chapter 1

Introduction

Just as *water utility system* varies in definition, so does *automation and instrumentation*. However, to provide a framework for this manual the following definitions will be used (AwwaRF/JWWA 1994):

Automation: the replacement or elimination of intermediate components of a system or steps in a process, especially those involving human intervention or decision making, by technologically more advanced ones.

Instrumentation: both the technology and installation of equipment to monitor and control operations and carry out information processing associated with observation or adjustments of operations.

In the broadest sense, an instrument is defined as a device that performs a specific job. In a water utility, an instrument is usually a measuring or control device. In an automatic system, the controlling factor, such as flow or pressure, has to be reliably sensed or measured. Automation and instrumentation are closely associated because one depends on the other.

THE WATER UTILITY SYSTEM

To provide a consistent approach, the following paragraphs apply to water treatment and distribution systems, their important elements, the operator's responsibilities, and automation and instrumentation's role.

A water distribution system delivers potable water, at a suitable pressure, in the amount required at customer service connections, through a piping network. The distribution system can consist of elements such as main pumping stations, booster pumping stations, storage reservoirs, standpipes, elevated tanks, water mains, valve stations, and wells. The operator has the duty to maintain the elements of the system and to see that they perform correctly and reliably.

An operator's main responsibilities are supervision and control. *Supervision* means examining system performance information and deciding if it is acceptable. If, in the operator's opinion, performance is unacceptable, then the operator must change an element or make an adjustment to the system to bring performance back to an acceptable condition. This is called *manual control*. When instruments are provided to make the necessary change or correction without the intervention of the operator, the system is called *automatic control*. However, regardless of the extent to which automatic control is used, the operator still may need to intervene manually during abnormal or emergency situations.

Because treatment plants, distribution system pumping stations, storage reservoirs, and other facilities may be at various, separate locations, the information needed to supervise and control the system must be gathered at some centralized point near the operator. Provision must also be made at this central location for remote control of any of the facilities that the operator may be required to regulate or change.

The operator will usually be working through some intermediate or intervening instrument to cause the systems to perform. Some of the instruments will be entirely mechanical, such as levers, chains, and cables; some will be hydraulic systems, using water or oil pressure for power sources and control; some will be pneumatic systems, using compressed air for power, control, and instrumentation; and some will be electrical systems for power, control, and instrumentation. A swing check valve will close automatically, for example, when not forced open by the flow of water through it. An indicating pointer can be positioned by a system of cable and pulleys to provide position indication. Oil or water pressure can be used to hold a valve closed, whereupon the loss of pressure will cause it to open automatically. Similarly, compressed-air pneumatic systems can be arranged to cause devices to operate automatically; pneumatic instrumentation and control systems are used extensively. Electricity is used more than any other source of power for control and instrumentation.

Generally, an electrical system, together with various mechanical, hydraulic, and pneumatic subsystems, allows an operator to supervise and control the water system. These electrical systems may include any or all of the following:

- Power system, using local, remote, or automatic control
- Telemetering, monitoring, and alarm system
- Communication system, data acquisition, and data processing

Operations are performed automatically for several reasons:

- the operator does not have to do them
- the operator cannot do them
- they can be done faster and better automatically
- they can be more efficient

As with automation, instrumentation is an extension of the operator. Instruments see, feel, measure, and record information for the operator. Instruments can perform a variety of operations, including:

- measuring
- monitoring
- comparing
- remembering
- signaling
- regulating
- calculating
- switching
- transmitting
- receiving
- recording
- indicating

- integrating
- timing
- converting
- summating
- anticipating
- detecting
- programming
- analyzing
- alarming

Each individual instrument is a single device with a specific task. Collectively, instrument systems can seem extremely complex; but the operator who understands each device and its unique function will be able to use each instrument as an aid to efficient supervision and control. An operator should be acquainted with all the automatic controls and instruments in use in the utility. This will give the operator the confidence needed to use the equipment effectively. With a broad knowledge of instrumentation and its applications, the operator becomes the driving force to seek operational improvement through the technology of instrumentation.

HOW TO USE THIS MANUAL

This manual introduces the major topics of automation and instrumentation. While not an exhaustive source of specific details, the manual can be used to identify water utility system automation and instrumentation elements. At a pumping station, for example, an operator will be able to examine an item of equipment and understand what it does, how it works, and identify the functions of its associated devices. Having identified a device, the operator may refer to the operations manual for further information. By using the plans of the station, equipment instruction manuals, equipment nameplate data, and other general information, the operator should be able to learn the names of the various devices and become acquainted with the intended purpose of an overall assembly of the equipment. This assembly can, for example, include a pump, motor, motor starter, and pump discharge valve, together with instrumentation and protective devices working together to perform a function as a complete unit.

This procedure can be used to learn more about local and remote controls, metering, and instrumentation. This manual can be used to understand their functions, as well as their relationships to the connected equipment.

This operator's manual may cover considerably more material than would apply to many small facilities; yet the manual may not mention every device found in a particular system. In these cases, the operator is encouraged to seek other references, some of which can be found in this manual.

The chapter arrangement of this manual is intended to group related topics. The second and third chapters review the hydraulic and electrical principles used in automation and instrumentation, as well as the basics of electric motor controls. The fourth and fifth chapters discuss instruments that measure process variables such as flow, pressure, level, and temperature. These types of instruments are called *primary instrumentation*. The sixth and seventh chapters present *secondary instrumentation*, those instruments that respond to and display information from primary instrumentation. The eighth chapter looks at the final control elements, such as pumps and valves. The last three chapters introduce the basics of automatic and digital control elements.

The topics of each chapter are introduced in the following paragraphs.

Chapter 2 Hydraulics and Electricity briefly reviews hydraulics and electrical power as the subjects relate to automation and instrumentation. Water utility operators are usually more familiar with hydraulics than electricity, and knowledge of both is necessary to get the most out of this manual.

Chapter 3 Motor Controls introduces the principles of the controls that stop and start motors, as well as the control of variable speed motors. The chapter also discusses motor control logic and presents motor control diagrams.

Chapter 4 Flowmeters discusses the most common flowmeters in service in water supply systems. These include the Venturi meter (Venturi), modified Venturis, orifice plate, magnetic, turbine and propeller, sonic, vortex, averaging Pitot, and rotameter. Also included are open channel flowmeters—weirs and flumes. Topics covered are basic theory, installation, maintenance, advantages, and disadvantages.

Chapter 5 Pressure, Level, Temperature, and Other Process Measurements introduces the primary sensors associated with three process variables encountered in water utility systems: pressure, level, and temperature. This chapter will touch on, in general terms, analytical instrumentation that is finding wide use in water systems, particularly in water treatment plants. An overview is also included of many of the less common sensors in use today.

Chapter 6 Secondary Instrumentation explains the pneumatic systems (those using air pressure) and electronic systems that control secondary instrumentation. Topics include the air supply system, pneumatic controllers, recording and indicating hardware, computing devices, converters, and applications.

Chapter 7 Telemetry is remote metering, taking a measurement at one location then transmitting it to another location. Specific topics include transmitting devices, output devices, controllers and function modules, communications, and various types of telemetry.

Chapter 8 Final Control Elements provides an overview for those applications that can produce a change in the process of treating and distributing water: valves and pumps. In general, this chapter describes the various types of final control elements and how they operate within an automated system.

Chapter 9 Basics of Automatic Process Control discusses how the elements presented in the previous chapters work together in a process that occurs without continuous operator input. The chapter provides basic information on process control and the most common techniques used to automate process control in water utilities.

Chapter 10 Digital Control and Communication Systems shows how computer and digital technology enable operators of process control systems to quickly recognize status changes and respond immediately. This chapter introduces the concepts, hardware, and software of digital control.

Chapter 11 Instrument Diagrams presents the standard instrument diagrams or process and instrument diagrams frequently used in the water utility systems.

REFERENCE

AwwaRF/JWWA (American Water Works Association Research Foundation/Japan Water Works Association). 1994. *Instrumentation & Computer Integration of Water Utility Operations.* Denver, Colo.: AwwaRF/AWWA.

AWWA MANUAL M2

Chapter 2

Hydraulics and Electricity

This chapter presents the basics of hydraulics and electricity. A basic knowledge of the physics of hydraulics (fluid mechanics) and electricity is required for any designer or operator to understand the system. Hydraulics topics include flows, pressures, elevation, valve positions, and other physical parameters that are sensed and transmitted for either monitoring or control. Topics on electricity include basic electricity, distribution concepts, power factor, and safety.

HYDRAULICS

In water systems, hydraulics explains how water acts in tanks, open channels, and pipes. The principles of hydraulics can be used to design sensing devices such as Pitot tubes, Venturi flow tubes, and ultrasonic, magnetic, and other static and dynamic sensing devices. Water, as a medium for transmitting force, can be applied to cylinders, pilot valves, and transmitter mechanisms. Hydraulics can be used to predict hydraulic surge and cavitation in order to properly design, control, and operate systems in a safe and efficient manner. Hydraulic friction in closed conduits and its relationship to flow directly determine the sizing of lines and valves, head losses, and the operation of different system configurations.

The three branches of hydraulics are

- Hydrostatics (liquids at rest)
- Hydrokinetics (liquids in motion)
- Hydrodynamics (forces exerted by or on a liquid in motion)

Properties of Liquids

Three basic characteristics of liquids are that

- they are virtually incompressible.
- they have unlimited directional movement.
- they can assume any form or shape.

Density of water. For most hydraulic equations, density, expressed as weight per unit volume, of pure water is assumed to be 62.4 lb/ft^3 (999.6 kg/m^3). This density is correct only at 52°F (11°C). As the temperature of water changes, its density changes.

- At higher temperatures water is less dense. At 200°F (93°C) density = 60.135 lb/ft^3 (963 kg/m^3).

- At lower temperatures water is more dense. At 39°F (3.9°C) density = 62.424 lb/ft^3 (1,000.03 kg/m^3).

- At 32°F (0°C) water freezes. Ice is less dense than liquid water; therefore, it floats.

Dissolved minerals and salts will add to the weight of water. The density of seawater is 64 lb/ft^3 (1,030 kg/m^3). Pure water at 39.2°F (4°C) is the standard used to determine the specific gravity of other liquids, assuming both are at atmospheric pressure, 14.7 psi (101 kPa). For these conditions, pure water is at its maximum density, 62.4266 lb/ft^3 (1,000 kg/m^3).

Incompressibility. Although liquids are shapeless, they are generally less compressible than a solid. When a force is applied to a confined liquid, the liquid acts in a manner very similar to a rigid solid. A pressure of 1 psi (6.9 kPa) will reduce a volume of water by 1 part in 300,000. For each drop of water added to a full container, a drop must come out or the container will burst. This is also true in a pipeline: a gallon in requires a gallon out.

Density and specific gravity. Pressure exerted by a liquid depends on its weight density. For example, water weighs 62.4 lb/ft^3 or 0.036 lb/in.^{3} (999.6 kg/m^3).

Viscosity. Viscosity is the internal frictional resistance a liquid offers to flow or motion. It increases with a decrease in temperature. Liquids such as polymers may have a viscosity that greatly affects the suction characteristics of chemical pumping equipment. Water is one of the least viscous liquids, whereas oil and molasses are more viscous.

Water used for backwashing a filter will have a different viscosity in the winter than in the summer. This affects the expansion of the filter bed. It would take 1,000 gpm (0.063 m^3/sec) at 86°F (30°C) but only 420 gpm (0.026 m^3/sec) at 32°F (0°C) to expand the bed equally.

Hydrostatics

Hydrostatics is the study of liquids at rest.

Hydrostatic pressure. Pressure, as the word is commonly used, means the intensity of force per unit area. Pascal's law states that pressure exerted at any point in a liquid acts equally in all directions and that this pressure acts at right angles to the surfaces of the containing vessel. As the example shown in Figure 2-1 illustrates, a 10-ft head in a tank is applied at the base of the tank in all directions, to the bottom and sides—downward and outward. The vertical distance between two horizontal levels in a liquid is defined as the head of liquid.

Every square inch of the bottom of the tank has the same head exerted on it. Similarly, at a depth of 5 ft, the pressure head is 5 ft in all directions.

The pressure at any point in a liquid depends on

- Height of liquid above the point

- Density of the liquid

- Any additional pressure applied to the surface

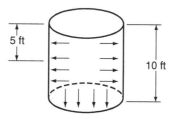

Figure 2-1 Pressure in a tank

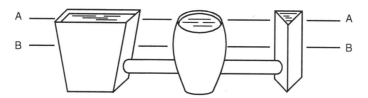

Figure 2-2 Pressure in containers of various shapes

All fluids will *seek their own level*; that is, the surfaces of any connected volumes of water will equalize at the same elevation. This is true even if two nonmiscible liquids (those that will not mix together in the same container) are placed on top of each other, such as water over mercury.

Effect of container shape on pressure. One of the corollaries of Pascal's law is that pressure is not altered by the physical shape of the container. To illustrate this, an arbitrary horizontal reference plane (called a datum plane) is used to measure hydraulic elevations, regardless of the container shape or pipeline grade. In Figure 2-2, with B–B as a datum plane, the pressure at B will be the same in each container; the same is true if A–A is used as the datum plane.

Hydraulic elevations, as referenced against the datum plane, can be positive or negative (above or below the datum plane). Sea level is commonly used as a reference plane.

Atmospheric pressure. All gases have weight, so the weight of air acting on the free surfaces of a liquid is another force to consider in evaluating hydraulic systems. Air weighs 0.075 lb/ft^3 (1.2 kg/m^3) at sea level.

Atmospheric pressure is

2,116 lb/ft^2 (101.3 kPa), or

(2,116 lb/ft^2)/(144 in.^{2}/ft^2) = 14.7 lb/in.^{2}, or

(2,116 lb/ft^2)/(62.4 lb/ft^3) = 33.9 ft of water, or

(2,116 lb/ft^2)/(848.6 lb/ft^3)/(12 in./ft) = 29.9 in. of mercury.

Vacuum. A perfect vacuum is a space containing nothing: no solids, no liquids, and no gases. A perfect vacuum is unattainable, and the amount of vacuum is equal to the amount that the pressure is below atmospheric pressure.

Liquids below atmospheric pressures obey Pascal's law just as they do at positive pressures. Negative pressures, referred to as negative heads, will be found in devices such as pumps, eductors, and Venturi tubes. Sensed pressures, both

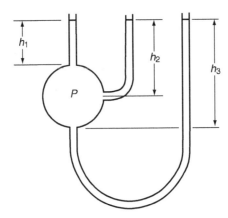

Figure 2-3   Water level in an unpressurized system

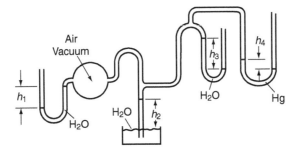

Figure 2-4   Fluid levels in a vacuum system

negative and positive, provide information, which is converted into flows, elevations, and pressures necessary to operate a water utility. The methods of conversion and instrumentation used in the process will be discussed later in this manual.

Summary of hydrostatics. The hydraulic arrangements in Figures 2-3 and 2-4 illustrate the basics of Pascal's law, under both positive and negative pressures, for a liquid under static conditions. The use of piezometric tubes can show the pressure at any point in a hydraulic system (see Figure 2-7). The figure shows that no matter where a group of piezometric tubes are connected, if cut through the same plane, the water level will be the same in every tube.

Similarly in Figure 2-4, a vacuum can be pulled on a system such that if h_1 = 12 in. (120 mm), h_2 and h_3 will also equal 12 in.

Hydrokinetics

Hydrokinetics describes the characteristics of liquids in motion.

Liquid flow. Some of the elementary characteristics of flow are volume, velocity, laminar flow, turbulent flow, and the energy involved.

Volume is the quantity of liquid that passes a known point in a hydraulic system, and velocity is the rate or speed of the liquid passing that point.

The velocity of the liquid flow increases as the cross-sectional area decreases. Also, the velocity decreases as the cross-sectional area increases (Figure 2-5). The flow velocity at B is two times that at A, because the area at B is half that at A; therefore, a constant flow is maintained throughout the length of the pipe.

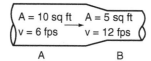

Figure 2-5 Flow velocity as a function of cross-sectional area

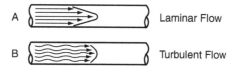

Figure 2-6 Flow–velocity profiles

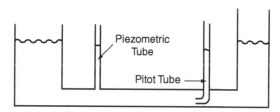

Figure 2-7 Determination of static pressure

Laminar and turbulent flow. If the average velocity is low, the flow of the liquid will be laminar; i.e., all the particles move parallel to each other without interaction. As flow rate increases, these particle streams will continue to run parallel until some critical velocity occurs, when the streams will waver and suddenly break up into turbulent flow.

Turbulent flow produces random motion of the liquid, even in a direction at right angles to the flow. The velocity distribution in the turbulent region is more uniform. However, even under these conditions, a thin layer of liquid at the pipe wall moves in laminar flow. If the velocity profiles of laminar and turbulent flow are measured and plotted, they will look something like those in Figure 2-6.

Measurements. Two conditions of flow in a closed conduit can be easily determined: static pressure and relative velocity. First, the static pressure at any point in the system can be determined with a piezometric tube (Figure 2-7) perpendicular to the pipe flow direction. The measurement should be taken at a point where the direction of the flowing fluid is not changing. Second, the relative velocity within the pipe can be measured using a Pitot tube. This device, with its tapered tube pointing into the flow, measures the velocity only at its opening. It does not, therefore, find the true mean velocity of the velocity profile that exists within a pipe, as shown in Figure 2-6. Several readings must be taken across the pipe at prescribed locations in a plane (a traverse) to calculate the average (or mean) velocity.

Energy and head. Water flow has potential energy (capacity to do work) because of its elevation, velocity, pressure, or any combination of these. Each energy form may be expressed as equivalent pressure or head in feet (or meters), or in pounds per square inch (or kilopascals). Also, each energy form can be converted to

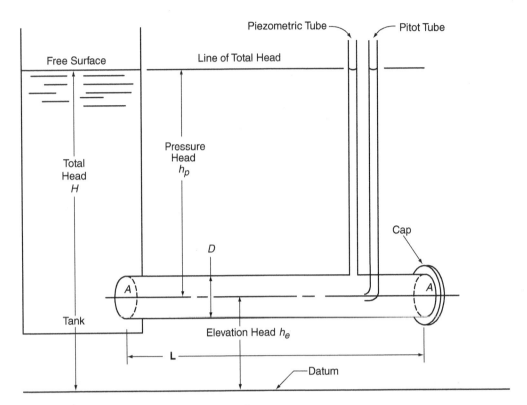

Figure 2-8 Water in pipe with pressure, no flow

the other two. The sum of these three energy forms is called the total head, H, (Figure 2-8). If water flowed through a smooth pipe without friction, the total head would remain the same at any section of the pipe (Figure 2-9). This would be true even if the diameter (D) of the pipe or its elevation changed.

Elevation head: h_e. In hydraulics, differences in elevation are measured above a selected datum plane. The pipe in Figure 2-9 is at the same elevation above the datum at all points; therefore, h_e is constant. If point 2 were higher than point 1, as in Figure 2-10, h_e at point 2 would increase and pressure head, h_p, at point 2 would decrease because part of h_p would be used to lift the water at point 2, where the energy would show an increase in h_e.

Velocity head: h_v. When water flows in a pipe, a part of the total head, H, is converted to kinetic energy (energy of motion) capable of lifting the water through a height, h_v, equal to the difference in elevation, h_e, if the water and pipe had no friction. This can be seen by the difference in levels in a Pitot tube, which senses velocity head (kinetic energy), and a piezometric tube, which senses static head energy (potential energy) at the point of measurement (Figure 2-11).

A body falling freely through a height, h, will attain a velocity expressed as:

$$v = \sqrt{2gh} \qquad (2\text{-}1)$$

Where:

v = velocity, theoretical, ft/sec (m/sec)
g = acceleration constant of gravity, 32.16 ft/sec^2 (9.8 m/sec^2)
h = vertical height or head, ft (m)

HYDRAULICS AND ELECTRICITY 11

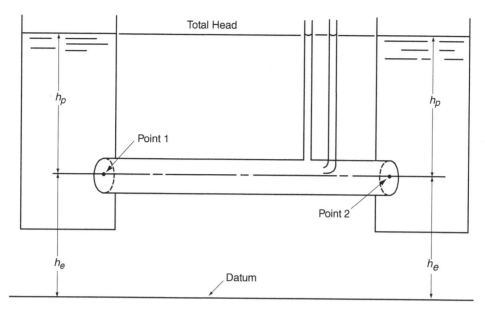

Figure 2-9 Total head

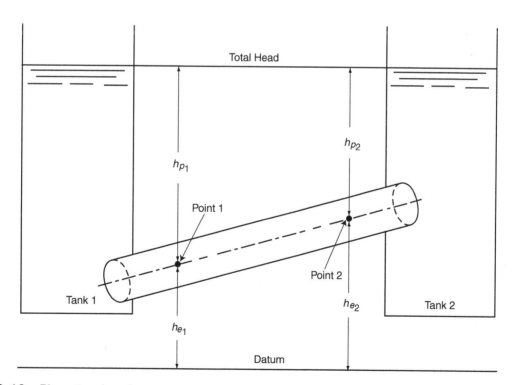

Figure 2-10 Elevation head

12 INSTRUMENTATION AND CONTROL

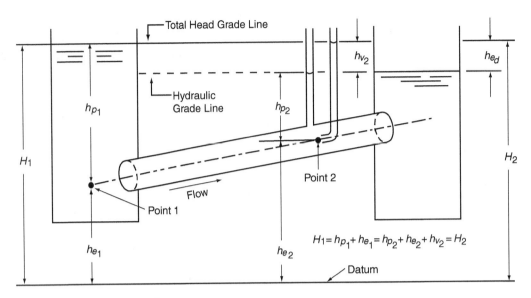

Figure 2-11 Flowing without friction

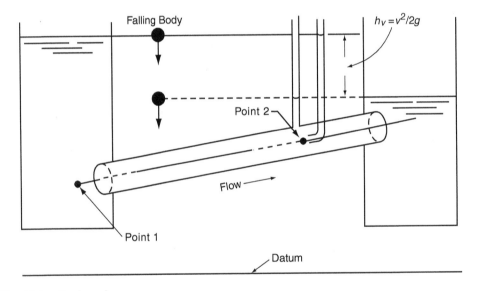

Figure 2-12 Velocity head

Water flowing through a pipe without friction is essentially falling through a height equal to the difference in elevations between two points (Figure 2-12). Consequently, the water in the pipe attains the same velocity as a falling body through a height equal to the velocity head, h_v, and this may be expressed as:

$$v = \sqrt{2gh} \qquad (2\text{-}2)$$

If a Pitot tube is inserted into the pipe, the energy of velocity will lift the water in the tube to a height equal to the difference in elevations, or to the total head minus the sum of the pressure head, h_p, and elevation head, h_e, or the head, h_v, causing the velocity. Consequently, h_v is called velocity head (represents kinetic energy) and can be expressed as:

$$h_v = v^2/2g \tag{2-3}$$

Pressure head: h_p. Pressure head, h_p, is equivalent to the pressure per unit area exerted against the walls of the pipe. It is the only part of the total head, H, that can be measured by a piezometric tube, manometer, or pressure gauge (and represents potential energy).

Friction head: h_f. When water flows from one point to another, turbulence, pipe roughness, and the frictional forces within the fluid cause friction, which generates heat, and therefore, head loss. Energy is not lost, because energy can be neither created nor destroyed; the energy is converted to heat.

Therefore, the rules for flow indicate that when water flows from one point to another, the sum of the elevation, velocity, and pressure heads at the second point must be equal to the total head at the first point, minus the friction head:

$$h_e + h_v + h_p = H - h_f \tag{2-4}$$

In Figure 2-13, because h_f is a continuously increasing value between point 1 and point 2, pressure head, h_p, decreases continuously between these same points. Because velocity head, h_v, is constant, the total-head hydraulic grade slopes in the magnitude of the increase of h_f while h_p decreases.

Total head: H. The combined effects of head can be summarized as follows: total head, H_1, at any point in the system is the sum of the elevation head, h_e, pressure head, h_p, and velocity head, h_v, at that point. The total head, H_2, at a second

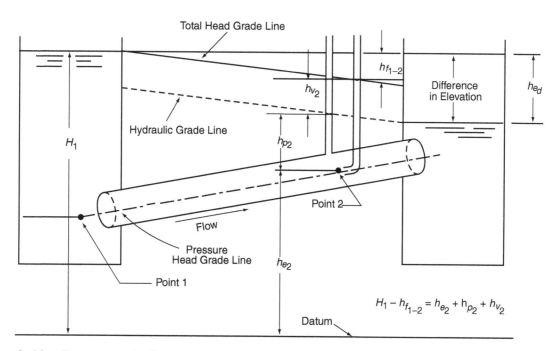

Figure 2-13 Flowing with friction

point, equals the total head, H_1, at the first point, minus the head loss, h_f, caused by friction between the points:

$$H_2 = H_1 - h_f \qquad (2\text{-}5)$$

Quantity flowing in straight pipe—no friction. If the flow is zero, water will stand at the line of total head in both the piezometric tube and the Pitot tube. No velocity head, h_v, exists. The volume, V, contained in length of pipe, L, between two points is determined as follows:

$$V = A \times L \qquad (2\text{-}6)$$

Where:

V = Total volume, in ft^3 (m^3)
A = Area of pipe cross section, in ft^2 (m^2)
L = Length of section, in ft (m)

If water flowed without friction, water in the piezometric tube would stand at a height, h_v, that was a distance below the line of total head, H, because that part of H is converted to velocity head, h_v, producing velocity, v.

As explained in the previous section, velocity, v, will be the same at all sections of a pipe of constant cross-sectional area. The hydraulic grade line drawn through the water elevation in a series of piezometric tubes will be parallel to the total head grade line.

The quantity, Q, in cubic feet (or cubic meters) for length, L, will flow past a point in a time period, t. Therefore,

$$L/t = \text{Velocity}, v, \text{ in ft/sec (m/sec)} \qquad (2\text{-}7)$$

Where:

t = time, in seconds

Therefore, the equation for still water in pipe, $Q = A \times L$, is changed to the following for flowing water:

$$Q_t = A \times v \qquad (2\text{-}8)$$

Where:

Q_t is equal to volume rate of flow in ft^3/sec (m^3/sec)

However, velocity is produced by the conversion of part of the total head, H, to velocity head, h_v. Velocity, v, is expressed as follows:

$$v = \sqrt{2gh_v} \qquad (2\text{-}9)$$

Substitution yields:

$$Q_t = A \times \sqrt{2gh_v} \qquad (2\text{-}10)$$

Quantity flowing in straight pipe—with friction. Water flows in the presence of friction. A part of the total head, H, is converted to velocity head, h_v, and a part of the total head, H, is lost in friction, h_f (Figure 2-14).

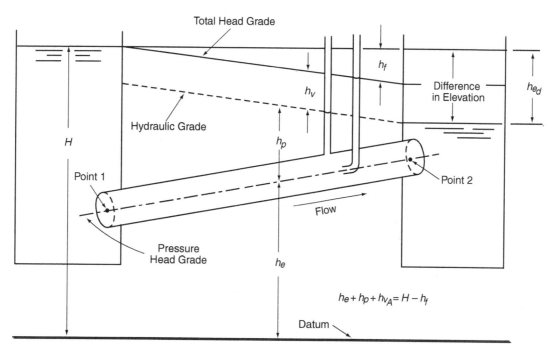

Figure 2-14 Flowing with friction

The value of friction, h_f, increases with distance between any two points. The pressure head, h_p, is decreased by the same amount that friction head, h_f, increases. Because the total head, H, includes the pressure head, the total head is decreased in the same amount. The water elevation in a series of piezometric tubes will stand below the total-head grade, and the hydraulic grade line will slope.

The velocity head, h_v, discharged through an orifice is decreased by a small amount (usually 1–2 percent) equal to the head loss caused by the orifice contraction, h_f, at the discharge point.

$$h_{v\ total} = h_v - h_{f\ discharge\ orifice} \qquad (2\text{-}11)$$

For a given total head, H, and a given head condition at the discharge point, the actual velocity, v_A, and velocity head, h_{vA}, will be less than the theoretical velocity, v, and velocity head, h_v. However, because the cross-sectional area of the pipe is constant, the actual velocity head, h_{vA}, and the actual velocity, v_A, will be constant.

The actual quantity, Q_A, corresponding to the actual velocity head, h_{vA}, can be expressed as:

$$Q_A = A \times \sqrt{2gh_{vA}} \qquad (2\text{-}12)$$

However, h_{vA} is some unknown part of h_v, because $h_{vA} = h_v - h_f$, and the value of h_f is unknown.

The value of h_f varies with length and diameter of pipe, and the surface characteristics of roughness of the pipe. The head loss in friction can be expressed:

$$h_f = f \times \frac{L}{D} \times \frac{v_A^2}{2g} \qquad (2\text{-}13)$$

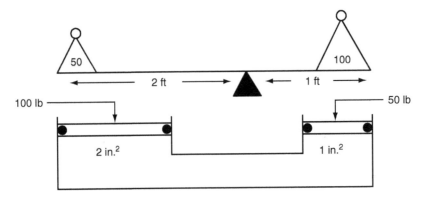

Figure 2-15 Mechanical leverage compared to hydraulic force

Where:
D = pipe diameter
f = the friction factor for the pipe in question

Combining the last two equations yields:

$$Q_A = A \times \sqrt{2g\left[h_v - \left(f \times \frac{L}{D} \times \frac{v_A^2}{2g}\right)\right]} \qquad (2\text{-}14)$$

This equation is not practical for determining actual quantity because too many unknowns exist: the theoretical velocity head, h_v, the friction factor, f, the actual pipe diameter, D, and the actual velocity, v_A. Therefore, to measure accurately the actual quantity, Q_A, some type of measuring device with known physical characteristics that can be bench tested must be used. Flowmeters serve this function to varying degrees.

Hydrodynamics

The forces of hydraulics are analogous to those of mechanical leverage, as shown in Figure 2-15 and discussed in the following paragraphs. The transmission of force in a hydraulic system is basic to chemical-pump equipment and cylinder-operated butterfly valves.

Force in hydraulic systems. Pascal's law states that a force applied to a confined liquid is transmitted equally in all directions throughout the liquid, regardless of the shape of the container. Transmission of pressure can be defined as force divided by the area over which it is distributed.

In Figure 2-16, 200 lb of force, F, applied to piston 1 will impart 20 psi to the liquid. Piston 2 will see the 20 psi and with its 10-in.^{2} area,

20 psi = F/10 in.^{2}

or 20 × 10 = 200 lb of force

Piston 3 will also see 20 psi; but with its 20-in.^{2} area, it will produce 400 lb of force:

20 psi = F/20 in.^{2}

or 400 lb of force

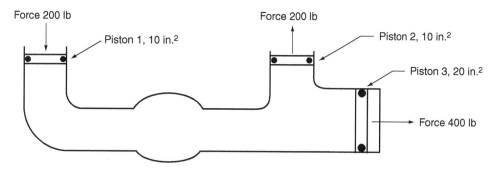

Figure 2-16 Hydraulic force

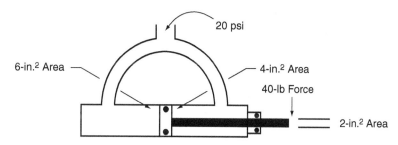

Figure 2-17 Differential areas

This multiplication of forces by the difference in area between two pistons is a principle commonly used in hydraulic equipment. Note that the shape of the connecting pipe has no effect on the forces.

The transmission of forces by hydraulic fluids can also be described as differential areas. In Figure 2-17, a single cylinder has a single piston rod extending through one of the cylinder end caps. The effective area on the left side of the piston is 6 in.^{2}. The effective area on the right side of the piston is 4 in.^{2}, because the rod occupies 2 in.^{2}. Consequently, there is a force of 120 lb on the left side of the piston and 80 lb on the right, leaving a net force of 40 lb on the end of the rod. The net effect is the result of the area differential created by the rod because all other effective areas balance out.

Energy and work. Work is defined as a force moving a mass through a distance, and the amount of work is the product of the force multiplied by the distance. Without friction, the work input will be equal to the work output.

Energy includes work, together with other forms of energy into which work may be converted, and the forms that can be converted into work. Work must involve motion, whereas energy can be at rest and still exist as energy. The two, work and energy, are always interchangeable. Because energy is never created or destroyed, the total energy output in all its forms will always exactly equal the total energy input; this is the law of conversion of energy. This law holds even if usable output does not equal the input. For example, the seal on a cylinder shaft has friction. Consequently, some of the work imparted to the cylinder is consumed by the friction, which, in turn, generates heat. The total energy input equals the usable output plus the heat energy resulting from friction.

ELECTRICITY

Electricity is the primary type of energy used to power the majority of equipment used in the transport, treatment, and distribution of water. Electricity is usually delivered to a single point in most facilities by the local electrical utility and then distributed within the facility using equipment owned and operated by the water utility.

This section is not intended to make the reader an expert in electrical systems, but rather to describe the basic principles and equipment used in electrical power distribution. Water utility operators will better understand how their facilities operate and their options for meeting different operating requirements.

This section explains the following:

- Basic Electricity—physical laws that govern how electricity behaves

- Distribution Concepts—basic techniques used in electrical power distribution and commonly used electrical diagrams

- Safety—basic safety issues and the equipment used to enhance safety in electrical circuits

- Power Factor—what power factor means, how it impacts efficiency, and how to improve it in a facility

- Lightning and Surge Protection—protecting people and equipment from lightning and other electrical surges

Basic Electricity

Electricity is the movement of electrical charge from one place to another. Electrical charge results from atoms that have more electrons than protons (a negative charge) or fewer electrons than protons (a positive charge). Positive charges always attract negative charges, and negative charges always attract positive charges. Conversely, positive charges always repel other positive charges, and negative charges always repel other negative charges. This attraction or repulsion is called an electrical force, which can be quite substantial. By manipulating these charges according to the laws of physics, electricity can be very useful.

All atoms have one or more protons in their center, or nucleus, which attract an equal number of electrons on the outside edge of the atom. Atoms which have an equal number of electrons and protons are electrically balanced; they have no charge. Because an atom's electrons are attracted to their protons by an electrical force, they have to be forced out of the atom to create an electrical charge. Electrons can be forced out of an electrically balanced atom in several ways. Each of these is briefly explained below:

Physical—Electrons are forced out of atoms colliding with nearby atoms. When atoms on the surface of one material collide with the atoms on the surface of the other material, some of the electrons bump into the electrons of the other atoms and move from one material to the other. When an electron is knocked out of an atom, the electron becomes negatively charged. Therefore, the atom becomes positively charged because it now has more protons than electrons.

The dislodged electron is called a *free electron*. A free electron remains free only if there is no atom with extra protons available to neutralize it. It can move through material without becoming attached to an atom. This kind of charge movement is called *static electricity*, which occurs in material that does not easily conduct electricity. One of the most dramatic forms of static electricity is lightning.

Thermal—Heat is a form of energy, and when atoms are heated, their movement increases. Heat can cause electrons to free themselves from atoms, becoming free electrons. This process works most effectively in materials that easily conduct electricity, like metals.

Magnetic—Magnetism is an electrical force that pushes or pulls charges. While magnetism does occur in nature, it is usually not strong enough to cause a noticeable amount of electricity. One exception is the aurora borealis or *northern lights*. The aurora borealis occurs near the North Pole where the earth's magnetic field is strongest. Strong magnetic fields can be created, and most electricity is made by passing a conductor through a magnetic field.

Chemical—Some chemical compositions (molecules) containing more than one type of atom can cause atoms to trade or give up electrons. This causes the atoms to become electrically charged, resulting in positively charged atoms and several free electrons. Corrosion and the electric charge found in batteries are examples of this process.

Photovoltaic—The energy in light rays can cause free electrons. Common examples include the photocells used to power small calculators, remote terminal units in supervisory control and data acquisition systems, and space satellites. The photovoltaic process occurs in materials that do not retain electrons very well. Special composite materials are made that perform more efficiently.

Conductors and insulators. Because any atom can be stripped of one or more of its electrons, any material can be a conductor of electricity. Materials are classified by how easily they conduct electricity. Three basic categories are used:

- Conductors—materials, usually metals (because they have more electrons than nonmetals), easily forced to carry electricity
- Semiconductors—materials easily forced to carry electricity only under certain circumstances
- Insulators—materials, usually nonmetals, that can only be forced to carry electricity by applying a very strong voltage

Physical laws. To predict how an electric charge will behave under certain conditions, it must be quantified. These electrical quantities are similar in concept to the common properties used to describe the movement of water from one place to another. As each electrical property is introduced below, its corresponding water flow quantity is also stated.

Because the amount of charge associated with one electron (or the absence of one electron) is so small (like a drop of water), a larger amount of charge, the coulomb, is used in measurement. A coulomb of electrical charge (similar to a gallon of water) is the charge of 6.25×10^{18} (6,250 million billion) individual electrons.

As water flow is the movement of water particles, electricity is the movement of charges. The velocity of electrical movement is the number of charges moving per unit of time. This velocity is called current, which is abbreviated by the letter I in formulas and measured in a unit called an ampere. Amperes are often referred to as amps. One ampere is equal to one coulomb moving past a point in one second, which is similar to the flow of water measured in gallons (or liters) per second.

The velocity of a charge is determined by the amount of force pushing it and by the type of material it is moving through. The force that pushes an electrical charge is called an electromotive force, or EMF. EMF is abbreviated by the letter E in formulas and is measured in volts. E is also referred to as *electrical potential* or simply *potential*. In a water system, the force that pushes water through a pipe is called the head and is measured in feet of water column. Often the amount of EMF used is several thousand volts; a kilovolt (kV) is one thousand volts.

The material a charge moves through will provide some resistance (R in formulas) and is measured in ohms. Resistance is similar to the coefficient of friction in a pipe. Conductors have low values of resistance, while insulators have high values of resistance.

These three properties of electricity (EMF, current, and resistance) are related by the formula called *Ohm's law*, which states that one volt of EMF will cause one ampere of current to flow through one ohm of resistance. Written mathematically, it would be:

$$E = I \times R \qquad (2\text{-}15)$$

Therefore, if any two quantities are known, it is simple to calculate the third quantity.

These electrical properties are popularly known by their units of measurement. That is, EMF is called *voltage*; current is called *amperage*; and resistance is called *ohmage*. All electrical properties use metric units. All electrical units are named for the name of the scientist that first discovered and defined the electrical property. For example, the EMF unit, volt, was named for the Italian scientist, Volta, whose early experiments with frog legs led to the discovery of EMF.

Another electrical quantity is power, P, which is a measure of the amount of electrical energy needed per unit of time to cause a specific current to flow through a specific resistance. Power is measured in units called *watts*. As with voltage, large amounts of power are measured in kilowatts (kW), equal to 1,000 watts. Electrical power measured in watts can be calculated by multiplying the voltage by the amperage as shown in the following formula:

$$P = E \times I = I^2 R = \frac{V^2}{R} \qquad (2\text{-}16)$$

or:

$$1 \text{ watt} = 1 \text{ volt} \times 1 \text{ ampere} \qquad (2\text{-}17)$$

A joule is the metric unit of energy. One watt is defined as one joule of energy flowing past a point in one second. The English unit of power is a horsepower, and the English unit of energy is a calorie. Calories are related to joules by the following equation:

$$1 \text{ calorie} = 4.18 \text{ joules} \qquad (2\text{-}18)$$

while horsepower is related to watts by the following equation:

$$\begin{aligned} 1 \text{ horsepower} &= 746 \text{ watts} \\ 1 \text{ hp} &= 550 \text{ ft-lb/sec} \end{aligned} \qquad (2\text{-}19)$$

Most electrical motors are rated in horsepower, so converting between the metric and the US customary unit systems of measurement is not necessary. However, a motor's rating is the amount of mechanical power that it delivers and not the electrical power that is required to make it run. Because of the power factor and motor inefficiencies, approximately 1,000 watts (1 kW) is required to power each horsepower of a motor.

Other factors. In addition to quantifiable factors of electricity, other factors explain electrical power distribution systems. One factor is that the system must have a continuous path for the current to flow from the source of power to the load where the work is performed and back to the source. This complete path is called a circuit.

Another factor is that electricity can be easily created in two fundamental forms—direct current (DC) and alternating current (AC). Direct current is the flow of charge in one direction only. In alternating current, a charge flows in one direction, stops, then flows in the other direction, then stops. This cycle is repeated endlessly until the circuit is broken.

The number of times this cycle is repeated in a specific time period is called the frequency. Frequency is measured in hertz, the number of cycles that occur in one second. Both DC and AC are useful, but because of safety and the ability to change voltages more easily, AC is most commonly used for power distribution applications.

These few electrical parameters and their relationships to each other are all that is necessary to understand the basics of electrical power distribution systems as presented in the remainder of this chapter.

Distribution Concepts

Electrical energy needs to be delivered from the place where it is made to the place where it is to be used. In remote locations, electricity may be generated by small generators or stored in batteries. However, for most sites, electricity is generated by an electric utility and delivered through wires to individual customer sites similar to the way water is distributed through pipes to customers by the water utility. In fact, most of the concepts of electrical distribution are the same as the concepts of water distribution. Two major differences make electricity slightly more complicated: electricity requires a return path and a higher degree of safety awareness than water. The return path requires another wire, whereas water distribution uses only one pipe.

Electricity is very dangerous. Very small amounts can kill a person, and if a large amount of electricity is pushed through a small conductor, it can cause a fire. For this reason, electricity distribution is strictly controlled by several rules developed by the National Fire Protection Agency (NFPA) and published in the National Electric Code (NEC). In the US, it is illegal to violate any rule in the NEC. Some jurisdictions amend the NEC with more stringent rules. Therefore, changes should not be made to an electrical distribution system by anyone who is not knowledgeable of the NEC and any local amendments. The NEC is updated and published every three years; always refer to the latest edition when checking current requirements.

Because the electrical power distribution within the water utility's facilities requires using equipment owned and operated by the water utility, electricity must be distributed from the point of connection to the electric utility to the individual pieces of equipment used within the facility. Understanding electrical distribution systems is aided by using diagrams. These diagrams are similar to water system schematics and are called *one-line diagrams* or *single-line diagrams*, even though a circuit may have as many as four wires. Each line on the diagram shows the electrical path from the source to the load. The specifics of these paths are unique to each facility but follow some general concepts.

Each piece of electrical distribution equipment has a standard symbol used in one-line diagrams. As each piece of equipment is introduced and described, the current standard symbol will be shown and a *generic* single-line diagram will be developed step by step. When single-line drawings are made during the design stage of a facility, a map legend is always made to define the symbols used in that set. Because the symbols vary, the legend of a drawing should be referred to.

Conductors. Wires are the primary distribution path for electricity, and some information about wires is needed to understand distribution systems. Wires are made out of metal because it is a good conductor of electricity. Copper and aluminum are commonly used metals; copper is a better conductor than aluminum, but it is more expensive, heavier, and more rigid than aluminum. Aluminum is difficult to terminate properly and requires special techniques to ensure terminations do not overheat. Because the disadvantages of copper are more significant when the conductors need to be large, aluminum is typically used only for larger wire sizes.

Smaller wire sizes are standardized using a system called the American Wire Gauge (AWG). The AWG system has sizes numbered from 1 to 40, with 40 being the smallest. Electrical power distribution is limited to wire sizes 18 and larger. The smaller sizes are used for smaller electronic applications. After establishing the AWG, wire larger than No. 1 became necessary, so the next larger size was called 0 or *one-ought*. Larger sizes added 0s, so 00 (*two-ought*), 000 (*three-ought*), and 0000 (*four-ought*) conductor sizes were adopted. When wire larger than AWG 0000 is needed, the wire is sized by its cross-sectional area measured in thousands of circular mils (MCM). A circular mil is the area of a circle with a diameter of one mil (one thousandth of an inch). Common sizes of large wire are 200 MCM, 350 MCM, 500 MCM, 750 MCM, and 1,000 MCM.

Electrical wires must be insulated from their surroundings while they are carrying electricity. The insulating material used depends on the application, ambient temperature, likelihood of contact with moisture, oil, or other chemicals, voltage level to be insulated, cost, and other factors.

The size of the wire, its material of construction, and the material used to insulate it determine how much current a wire can safely handle. This capacity is called the ampacity of the wire. Ampacities of the various sizes of wire in copper and aluminum are in the NEC for several types of insulating materials.

Ensuring that too much electricity does not flow through a given conductor is critical to distributing electricity. Overloading a conductor may cause a fire or the breakdown of insulating material, allowing the electricity to escape and become dangerous. For this reason, all approved conductors of electricity, such as wires, switches, and transformers, have maximum current ratings. To ensure that too much current does not flow through the device, a current-limiting device, such as a fuse or circuit breaker, must be used. Current-limiting devices must always be used and *must never be replaced with devices having a higher rating than originally installed or shown on the design drawings for the facility*. If there is ANY doubt about the correct size to use, a qualified electrician or electrical engineer should be consulted. Because these devices are designed to protect conductors from abnormally high currents, they will blow or trip if something is wrong. A qualified electrician should be consulted before replacing fuses or resetting circuit breakers.

Utility service connection. Electricity is typically delivered to a facility at a single point. Wires may come from the utility on electrical poles or underground, usually at a high voltage. The amount of voltage depends on the amount of electricity that the facility needs; the higher the amount of energy, the higher the delivery voltage. Standard voltage levels vary from utility to utility, but some common values are 480 volts, 4,160 volts, 13,800 volts, 21,000 volts, and 69,000 volts. This energy is metered by the electric utility to measure watt-hours (the amount of energy used). The meters may also measure watts of demand (the highest level of power used at one time) and may include a clock when time-of-day billing is used. A switch is usually located on the upstream side of the meter and on the downstream side of the meter to allow the electric utility to service the meter or disconnect from their system for maintenance. These switches are called *disconnect switches* or just *disconnects*. The downstream side of the meter terminals or disconnects is the usual place for customer connections. Because this equipment is owned and operated by the electric utility company, it is usually kept inside a locked enclosure.

Voltage levels. In general, the goal of electrical power distribution is to deliver the required power to the equipment. For example, a current of 30 amps at 120 volts will deliver 3,600 watts, and a current of 15 amps at 240 volts will also deliver 3,600 watts. Is one better than the other? The answer is *yes* because larger currents require larger, more expensive wires. However, to insulate against higher voltages requires thicker, more expensive insulation. The deciding factor is that the power lost in the wire on the way to the equipment (similar to head loss in a pipe) depends on the resistance of the wire and the current flowing in the wire and not the voltage. Using the power equation, $P = I \times E$, and substituting the formula for E as given by Ohm's law, $E = I \times R$, then the power equation becomes:

$$P = I \times (I \times R) = I^2 \times R \tag{2-20}$$

If the resistance of the wire for R is used in the equation, the power lost to heat from the electricity flowing through the wire is the result.

If the current, I, doubles (to get more power) and the resistance is cut in half (to keep the voltage drop consistent) by using wire that is twice as big, then the power loss will be twice as much. Therefore, it is less expensive to increase the voltage and not the current for larger amounts of power. Utilities usually transport large amounts of power at high voltages.

Three-phase power. Another technique used to carry current to deliver more power not requiring the increase of current or voltage is to use three wires (phases) instead of two. Three-phase power only works with AC and cannot be used with DC. Three-phase power saves both the energy loss associated with increasing the current and the insulation cost associated with increasing the voltage. Seventy-three percent more power can be sent with only a 50 percent increase in wire costs. For small loads, however, the reduction in wire size using three-phase power results in wire so small that it is easily broken and therefore not very practical. For this reason, three-phase AC power is normally used only for loads larger than one-half to three-fourths horsepower.

Voltage conversion. Because electricity is usually delivered at a voltage much higher than needed, it must be converted or *stepped down*. This conversion is performed by a transformer. Transformers may be used to step voltage down or up but are most often used to step voltage down. Transformers may also be used to convert currents and are often used to reduce currents to permit sensitive instruments to measure large power conductor currents. These are called current transformers, sometimes referred to as CTs. Transformers have coils of wires called *windings*. They contain two sets of windings, one connected to the source of power (the primary winding) and one connected to the load (the secondary winding). The symbol for a transformer is shown in Figure 2-18.

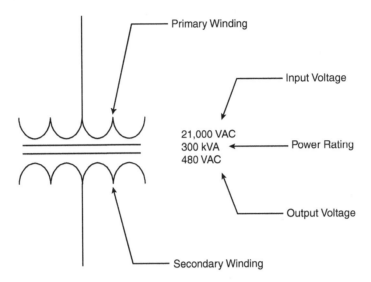

Figure 2-18 Transformer symbol

Transformers produce heat and larger ones require special treatment to withstand the heat. Excessive heat causes the insulating material between the windings to break down and eventually short the windings together. To accommodate higher temperatures, windings are commonly immersed in oil. The oil acts as the insulator between the coils of the windings and permits the transformer to run hotter than dry types of insulation. Transformers that do not use oil are referred to as *dry-type* transformers. Because oil-filled transformers are fire hazards, they are not permitted indoors. Additional cooling can be provided for either dry-type or oil-filled transformers by adding fans to blow air across them.

If the electric utility delivers the energy at the same voltage level at which it will be used (for example, 480 volts), then a step-down transformer is provided by the electric utility, although the transformer may be located on the water utility's property. These transformers are often located inside a fenced enclosure locked by the electric utility company. In larger facilities where the delivered voltage is quite high (such as 21,000 volts or higher), the voltage may be stepped down to an intermediate level, typically 4,160 volts (2,300 volts is common in older facilities).

Switches, enclosures, and current-limiting devices provide disconnection and safety protection for transformer installations, and all of these together are called a *substation*. Generic single-line diagrams showing a metering facility and substation with electric-utility-owned transformer and with water-utility-owned transformer are shown in Figures 2-19 (a & b).

Distribution. After the voltage is reduced to a level suitable for in-plant distribution, the electricity is directed to a bank of switches called *switchgear*. Switchgear directs the electricity to several different areas in the facility where it can be further stepped down in voltage or distributed to several loads in the area. These remote areas are called *load centers*. Switches allow each area to be isolated in case of a problem without affecting other areas of the plant. These separate paths are called *branch circuits* or *branches*. Each branch has a switch with a current-limiting device. Figure 2-20 shows the generic single-line diagram with the switchgear added. A plant switchgear is often located in a locked room to prevent unsafe or unauthorized operation.

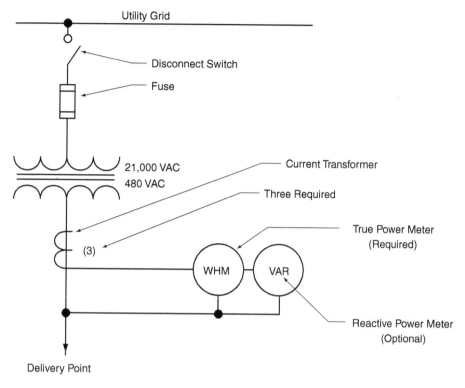

Figure 2-19a Delivery voltage at 480 VAC using electric utility's transformer

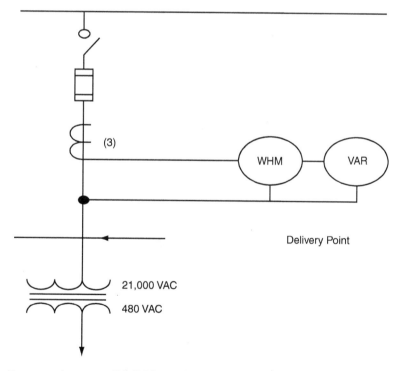

Figure 2-19b Delivery voltage at 21,000 VAC using water utility's transformer

26   INSTRUMENTATION AND CONTROL

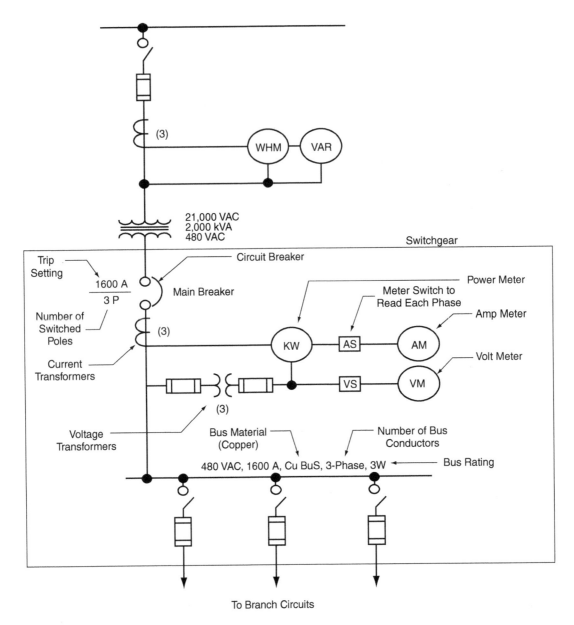

Figure 2-20 Main substation with switchgear

Load centers are also composed of switches. The switches control electricity to individual pieces of equipment or a group of small loads such as lighting. Load centers can have large switches arranged similarly to switchgear; they can also have special switches called *motor starters,* described in chapter 3, Motor Controls.

When several motor starters are installed in one location, the assembly is called a *motor control center,* or simply MCC. Load centers can also be a wall panel of small fused switches or circuit breakers. Load centers might also have substations on their upstream side to reduce the in-plant distribution voltage to a lower level before feeding the energy to the switches.

Equipment requiring more than one-half horsepower operates at 480 volts and smaller equipment at 208 volts or 120 volts, making it necessary to use transformers on the downstream side of a load center to reduce the voltage. When this occurs, the first load center is called a primary load center, and the smaller one is called a secondary load center. Often, these secondary load centers are small panels containing small circuit breakers, such as the ones used in residences. Figure 2-21 illustrates the generic single-line completed to show the electricity being delivered to load centers and on to individual pieces of equipment.

Functionally, the load center does exactly what the substation and switchgear for the plant do except for a smaller area and smaller amount of electricity. The number of load centers and their capacity is determined by the amount of electricity needed to be controlled and the amount and size of equipment requiring electricity.

Reliability. The reliability of the electrical power distribution system must be considered when any facility is built or modified. Reliability of electrical power distribution equipment is typically quite good, and electrical system failures are rare. However, when they do fail, the ability to treat or distribute water will stop. Therefore, if the process is critical, electrical systems will need some form of backup. Backup systems for electrical power distribution include having more than one source of power, more than one path for the electricity to access the load, or both.

A backup or secondary source can use power from the electric utility, a generator, or batteries. Which is appropriate depends on how much energy is needed. When large amounts of energy are needed, such as for large treatment plants, the second source from the electric utility is used. If the facility's capacity, and therefore the amount of energy, can be reduced in the backup mode or if the facility is small, a standby generator can be used to supply enough energy until the electric utility can re-energize the main feeder. For small critical loads, such as remote communications or plant exit lights, batteries are more economical.

For critical electronic loads, such as computers and instrumentation, an uninterruptible power supply (UPS) can be used. A UPS is composed of a rectifier, a battery, and an inverter. The rectifier is connected to the main source of AC electricity and converts the AC to DC, keeping the batteries charged. The batteries are connected to an inverter, which converts the DC energy in the batteries back to AC power. The output of the inverter is used to power the equipment. When the main power fails, the batteries continue to power the inverter, which in turn continues to power the critical loads. This prevents an interruption of power when the main power fails. UPSs can provide power to critical loads for a few minutes or up to several hours, depending on the size of the batteries.

Multiple paths are used within a facility to protect against failure of in-plant electrical distribution equipment. They can also provide alternative paths around equipment that is de-energized for maintenance purposes. Normally, only one extra path is provided. The extra path can include extra transformers, switches, and wires, depending on the importance of the loads. When a facility's electrical power distribution system has these extra paths, it has a *split* or *double-ended* architecture. As with secondary sources of electricity, these extra distribution paths may have as much capacity as the primary distribution paths or have reduced capacity depending on load importance.

Safety. Electricity is dangerous, and a small amount can be deadly. Electricity can cause fires, especially in areas where volatile chemicals, such as gasoline or paint, are stored. To help make electricity safe, electrical equipment has special devices and enclosures designed to prevent excessive heat and sparks and to prevent someone from touching a part of the system that carries current. Electrical systems

28 INSTRUMENTATION AND CONTROL

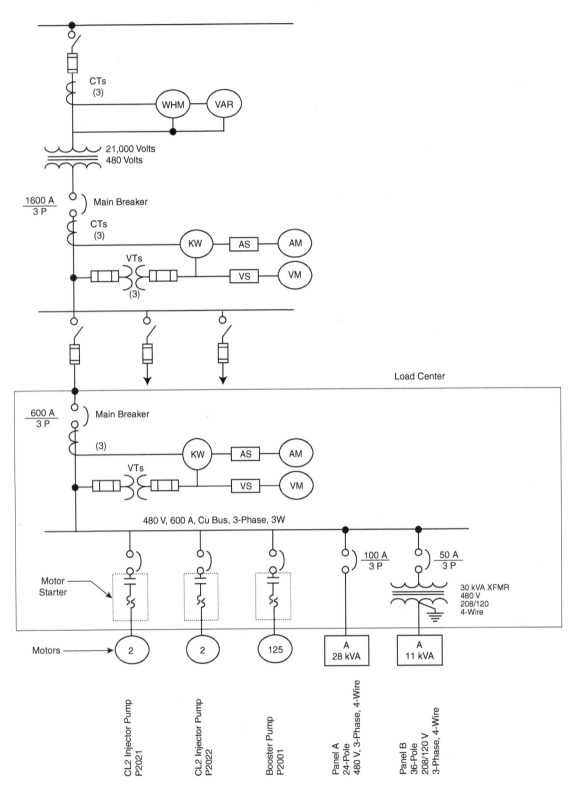

Figure 2-21 Complete one-line with load center and motors

should never be modified or opened except by a person trained in proper maintenance procedures.

Personnel hazards. A small amount of electricity can kill because the human body's nervous system uses small amounts of internal electricity to send the signals that control muscles and vital organs. A current as small as 20 milliamperes (20 thousandths of an ampere) can cause the heart or lungs to stop. A person must never touch any portion of an electrical distribution system that may be carrying current.

How an electric shock affects a person depends on whether the current is DC or AC, how much current flows through the person, and what part of the body it flows through. DC shocks are quite dramatic and cause the body to jerk. The electricity causes one or more muscles to contract or extend depending on the direction of current through the body. DC shocks usually disconnect the person from the current. In AC systems, the danger is much greater because the person cannot release the current source. AC current causes the muscles to contract and extend at the frequency of the current, at 60 hertz in power systems.

> DO NOT TOUCH a person who is being shocked, as the current will also shock you. The only way to help is to stop the electricity. If the switch cannot be located, the person should be pushed or pulled off the electrical equipment using a nonconducting material, such as dry wood, dry leather, or plastic. A stick can be used to push, while a belt or shirt can be removed and used like a rope to pull the person off the equipment.

After stopping the electricity from flowing through the person, call for emergency medical assistance, without leaving the victim alone. If the person is unconscious, check for heartbeat first, and then see if the victim is breathing. Standard first-aid techniques for cardiopulmonary resuscitation, or CPR, should be used to reinstate pulse and breathing until medical help arrives. The victim should be treated for shock and trauma by being laid down, covered to maintain warmth, and reassured. Electrical shock victims should always be checked by a doctor as soon as possible.

Fire hazards. Electricity causes fires by generating heat that occurs when current flows through a conductor. The amount of power lost or heat produced is proportional to the amount of resistance and the square of the current. Anything that carries electricity is going to create some heat. Problems occur when the amount of heat is enough to cause a fire. Electricity can also generate heat by arcing, which is a natural consequence of opening an electrical circuit while electricity is flowing. Because switches and other devices will open energized electrical circuits during normal operation, they must be designed to prevent arcing from igniting fires.

Wire sizes, insulating material, and switches that safely handle the required current without overheating should be used. This will prevent spontaneous combustion in the air around the conductors. The amount of heat that causes spontaneous combustion is called the lower explosive limit (LEL). The LEL is determined by the content of the air. For example, excess oxygen or volatile flammable chemicals, such as gasoline, lower the LEL and, therefore, lower the safe operating temperature. Consequently, the intended use of the area is considered when the electrical system is designed. To help standardize safe design practices for electrical distribution systems, the NEC has defined or classified areas by the degree of fire danger that they pose.

NEC area classifications have three categories. The first type is defined by whether flammable components will be present in the atmosphere of the area. The second classification is based on how flammable these components are. The last classification is based on whether the flammable components are dust, vapor, or gas.

During the design or modification of a facility, the proper area classification is determined, and the electrical system is designed to not exceed the safe operating temperature for that area. Therefore, if an area's use is changed from a lower level of fire danger to a higher level of fire danger, the electrical system must also be changed to ensure that normal operation cannot cause fires.

Fires caused during operation usually occur when more excess current flows in a portion of the system, or a lower LEL (more explosive) is present. Electrical systems are designed with several automatic methods to prevent fires from occurring, but these safety systems can fail.

Whenever an electrical fire does occur, the electricity should be shut off, a call for help should be made, and an effort made to extinguish the fire. Calling for help before trying to extinguish the fire is important; *do not let the fire block an exit route.* When fighting electrical fires, *do not use water if the electricity cannot be turned off. Water conducts electricity and electrocution may occur.* Use a fire extinguisher rated for electrical fires. It should be marked with a large, uppercase *C*. The extinguisher may also be marked with an *A* or a *B*, which indicates that it is suitable for other types of fires.

Safety systems. To minimize the danger of electricity, electrical distribution systems are designed to reduce the likelihood that electrical shock or excessive heat will occur. Because these safety systems are not foolproof, understanding the basics of these systems will maximize their effectiveness. Safety systems that are discussed in this section are

- Grounding
- Overcurrent and overtemperature
- Ground fault
- Encasement

While other protection systems exist, these are the most common and will have the greatest impact on safety.

Grounding. Electrical current will always take the path of least resistance back to its source; current can be diverted by offering a path of lower resistance. Electrical system design provides very low resistance paths back to the place where the electricity originates. These very low resistance paths are called the grounding systems. Their main purpose is to divert electricity away from places where people might inadvertently come into contact with the system.

The earth is an excellent conductor of electricity and is frequently used as part of the grounding system. When used for this purpose, the electrical grounds are made by driving a metal ground rod, sometimes called *earths*, several feet into the ground to get a good connection. In some locations, the soil is too dry or has too many insulating compounds and does not make a good ground conductor. In these locations, a grid or large mesh made of metallic wires is buried in the ground to make the earth a better conductor. These are called *ground grids*.

Grounding systems provide an easier path than human bodies, thereby protecting people against electrocution. The grounding system must never be circumvented. This is especially true in wet areas because water conducts electricity fairly well, especially dirty water. If someone is standing in water and touching an energized circuit and if the water is touching a grounded surface like a metal water pipe, electrocution may occur.

Portable power tools are often designed to be grounded by plugging them into a power receptacle with a three-prong plug. While many light-duty tools designed for home use have plastic cases and rely on double insulation to protect the user from electrocution, industrial-grade tools have metal cases and rely on the facility's grounding system to provide safe operation. If receptacles do not have the hole for the third prong, they are not grounded. *Therefore, do not use the tool with those receptacles.* Using a converter from three-prong to two-prong (called a *ground buster*) will circumvent the ground system and may cause electrocution.

As with industrial-grade power tools, most enclosures for electrical equipment and wires are made out of metal. Normally, all the current-carrying components inside these enclosures are insulated from the metal case so the case is not normally energized. However, if the insulation fails because of aging or mechanical failure, the normal current-carrying conductors can touch the case and cause the case to become energized. This is referred to as a *ground fault* or *grounding short*. To reduce this possibility, the cases are grounded, offering a low resistance path from the metal case back to the source of the electricity. This path is called the *ground path*, which is usually a wire connected from the case to the source of the electricity. The wire is referred to as the *equipment grounding conductor* and does not carry current during normal operation. Current flows in the wire only when the normal current path is inadvertently shorted to the case. The circuit will function normally if the grounding conductor is cut or broken; however, the protection against electrocution will be lost.

Overcurrent and overtemperature. Overcurrent and overtemperature systems are designed to prevent too much current from flowing in a conductor. Overcurrent can be detected by the rise in heat or the rise in magnetism. To protect against overcurrent, it must be detected and then the circuit must be de-energized until the cause of the overcurrent can be located and fixed. *Electrical circuits must never be operated without their overcurrent devices installed and operational, otherwise the risk of fire increases.* Two basic devices are used: fuses and circuit breakers.

Fuses are conductors made of special metals that melt at a temperature slightly higher than the temperature associated with the maximum expected current. If the current rises above normal, the fuse wire actually melts and opens the circuit. Once this happens, the fuse is useless and is said to be *blown*. To get the circuit working again, the cause of the overcurrent must be corrected and the fuse replaced.

Fuses are rated by the amount of current that they can carry without blowing and by the amount of time they take to blow at various overcurrent values. In general, they are available as both fast acting and delayed action. Fuses come in a wide variety of ratings and case styles. A replacement fuse must be one that has the same case style because fuse holders are designed to accept only one type.

Circuit breakers are switches designed to carry their rated current but will *trip* open if the rated current is exceeded. When a circuit breaker trips, the cause of the trip should be fixed and the breaker reclosed to place the circuit back in operation. Because they do not have to be replaced, they are less expensive to maintain than fuses. However, they are more expensive to buy and cannot be made with the precision of fuses. There are two mechanisms commonly used to open or trip the switch—thermal and magnetic. Circuit breakers are available with either or both mechanisms installed.

Thermal circuit breakers have a bimetal strip that will bend when heated. As the current increases above the breaker's rating, the strip bends until it trips the breaker. These circuit breakers operate relatively slowly and are comparable to delayed-action fuses.

Magnetic circuit breakers react to the strength of the magnetic field caused by the current flowing through them. When the current increases, it produces a stronger magnetic field, which increases the strength of a small electromagnet built into the switch. As the current continues to increase, eventually the electromagnet's strength will be adequate to trip the breaker and open the circuit. This type of breaker can open the circuit very fast if the amount of overcurrent is high but does not operate on small values of overcurrent.

Never replace fuses and circuit breakers with units having higher current ratings or different time characteristics. If you do, the protection will be less effective and the risk of having a fire or damaging equipment increases.

Ground fault. When ground faults occur, the circuit should be stopped until it can be repaired. Failure to do this may result in excessive ground currents causing excessive heat and possibly fires. If the ground fault is in a low-resistance path, a large amount of current will be diverted to ground, and the overcurrent devices will interrupt the circuit. However, if the ground fault is not in a low-resistance path (such as with a person's body), then only a small amount of current will be diverted, which will not be adequate to trip the overcurrent device. Additional safety procedures should be provided that will detect these small ground faults and interrupt the circuit. Ground fault detection systems or devices are designed to sense these small ground faults and interrupt the circuit to prevent accidental electrocution.

The wires used to conduct electricity typically include a grounding conductor. Because the grounding conductor does not normally conduct any current, the current that goes down one of the two circuit wires comes back on the other circuit wire. If the current on the two circuit wires is not equal, there must be another path for the current to flow in. This other path is not an intended path, indicating that something is wrong.

Ground fault detection systems measure all the currents in all circuit wires (including the neutral but not the ground), and if they differ by a small amount, a circuit breaker is forced to open and de-energize the circuit. By measuring the differential current in the main circuit wires and not the presence of current in the grounding conductors, even systems that do not have a grounding conductor (many older circuits do not) can be provided with effective ground fault protection systems.

Encasement. When electrical switches must be used in an area where explosive atmospheres may exist, the switch contacts must be encased to prevent sparks from igniting the atmosphere. This encasement is done in one of two ways.

Small switches can be manufactured in a way that completely encapsulates them in a sealed box. These are called *hermetically sealed contacts*. Because they are located in a sealed environment, the explosive atmosphere never comes into contact with the switch contacts and, therefore, cannot be ignited by the arc. A disadvantage of this approach is that the box cannot be opened for maintenance and the whole unit must be replaced if it fails.

Medium-size contacts are too expensive to encapsulate in hermetically sealed boxes, so they are put in enclosures that are rated explosion-proof. This means that even though the arc may ignite an explosive atmosphere inside the enclosure, the explosion cannot spread to the exterior of the enclosure. The box is made strong enough to withstand the internal explosion and has its openings designed to automatically extinguish flames. This latter feature is accomplished by making all the paths between the inside and outside of the box much longer than their cross-sectional area. Explosion-proof boxes have many threads on their screw covers to accommodate this design. Whenever an explosion occurs in one of these boxes, the equipment inside will typically be destroyed and is not suitable for areas in which the atmosphere is likely to be explosive.

The boxes are used in areas that might contain an explosive atmosphere if something were to break, for example, a room with fuel pipes passing through it. If the pipe breaks, then the room will have an explosive atmosphere.

Large electrical switches make large arcs that are difficult to contain and should not be used in an explosive environment. When it is necessary to locate them in areas that *may* become explosive, the area's LEL should be monitored. If the atmosphere becomes explosive, then the circuits should be de-energized. In these situations, the electrical system should not be used if the monitoring system is not operational.

Because electrical distribution systems are routed throughout a facility and typically use small pipes (called conduits) to carry the wires, they can transport explosive gases from a hazardous area to a nonhazardous area. This potential must be prevented by sealing the conduits with a compound that prevents any of the gas in the hazardous area from leaving the hazardous area.

Power Factor

The power factor of an electrical power distribution system has a significant impact on the efficiency of the system. The factor is a relative measure of how much of the energy that goes into a system is consumed in line losses versus productive work. A comparison to a water plant would be starting up a plant by pumping water into the plant and not producing treated water until all the tanks and pipes have filled. The energy used to fill up the storage volume is necessary to place the plant in service, but it does not produce any treated water.

DC electrical systems are similar in that they require a certain amount of energy to be stored before the charges actually get to the load where they are productive. However, with the water plant and DC electrical systems, the energy for operation soon becomes a much larger amount than the energy used to fill them, and the small amount of energy wasted in the beginning is not significant.

With DC electrical systems and the water plant examples, the current or water always flows in the same direction while the system is in operation. Unfortunately, AC systems actually stop operating in one direction, drain the system of charge, and then fill it up in the other direction. This occurs 60 times every second as long as the circuit is operating. Consequently, this fill-up energy continues to be needed as long as the system is operating.

Power factor is the ratio of the amount of energy used to operate the system to the total apparent amount in the system. Ideally, the power factor would be one, which would mean that all the energy that enters the system is consumed as line losses or productive work. Unfortunately, only DC electrical systems inherently have unity power factors. All real AC systems have a power factor less than one. To bring the power factor closer to one, that is, toward a more efficient system, the fill-up energy needs to be minimized. Understanding how to do this is discussed in the following section.

Reactance. The resistance of the electrical circuit and the amount of current flowing determines the amount of energy that will be lost. While this fact is true for DC and AC circuits, AC circuits are more complex. AC circuits also have reactance (X) that is measured in ohms. The total opposition to current flow in an AC electrical circuit is the result of resistance (R) and reactance (X). Reactance is what accounts for fill-up energy. The combination of resistance and reactance is called impedance (Z) because it impedes the flow of current. The equation for impedance is

$$Z = \sqrt{R^2 + X^2} \qquad (2\text{-}21)$$

Ohm's law for AC circuits becomes:

$$E = I \times Z \quad (2\text{-}22)$$

and the power equation becomes:

$$P = I^2 \times Z$$

This power is referred to as apparent power (P_A) because it is the amount of power that is apparently required to operate the circuit. Apparent power is measured in volt-amperes or volt-amps (VA). As with impedance, apparent power is composed of both reactive and resistive components. These components are called reactive power, which is measured in volt-amps-reactive (VAR), and true power, which is measured in watts. Reactive power (P_R) is the amount of power used up in filling the reactive part of the circuit. True power (P_T) is the amount of power used in the resistive part of the circuit. Apparent power is related to its two components in a similar manner as the impedance equation. The relationship is

$$P_A = \sqrt{P_T^2 + P_R^2} \quad (2\text{-}23)$$

The power factor is calculated by the formula:

$$\text{Power Factor} = P_T / P_A \quad (2\text{-}24)$$

To fully analyze what happens with voltage, current, and power in AC circuits is beyond the scope of this manual. However, reactance will be discussed to better explain how it affects the efficiency of electrical power distribution systems. Reactance is the combined effect of the electrical properties' inductance and capacitance.

Inductance. Inductance (L) is similar to inertia in a water distribution system, which keeps the current flow constant. To slow current down and reverse its flow as required in AC systems, inductance must be overcome. Inductance is measured in units called *henrys*. Every conductor has a small amount of inductance. When wires are coiled as they are in transformers or motor windings, the amount of inductance is significantly increased. Reactance caused by inductance is called inductive reactance (X_L) and is measured in ohms.

Capacitance. Capacitance (C), which is similar to water storage in a tank, keeps the voltage constant. To get voltage to increase and push current in one direction and then decline and build up in the other direction as required in AC systems, capacitance must be overcome. Capacitance is measured in units called *farads*. All electrical circuits have a small amount of capacitance between each conductor and between each conductor and ground. Reactance caused by capacitance is called capacitive reactance (X_C) and is measured in ohms.

These two reactances act in opposition to each other so the total circuit reactance is equal to the difference between them. The formula for inductance is

$$X = X_L - X_C \quad (2\text{-}25)$$

This equation indicates that if the circuit is too heavily inductive, it can be nullified by adding capacitance. Because power distribution circuits are inherently inductive due to the transformers and motors, capacitance needs to be added to bring the power factor closer to one. Adding capacitance can be done by installing capacitors or using capacitor motors. A brief discussion of each of these techniques follows.

Capacitors. Capacitors are devices that provide additional capacitance in electrical circuits. They are made in very small sizes for electronics and in very large sizes for power applications. Power capacitors are rated in the amount of VARs or kvars of reactive power that they consume at 60 hertz and the voltage they can safely withstand. They are connected to the circuit at points where there are high amounts of inductance so that they can neutralize the inductance. Capacitors can be in motors, in MCCs, or in the main switchgear.

Capacitive motors. Capacitive motors use capacitance instead of inductance to turn their rotors and, therefore, have capacitive reactance instead of inductive reactance. These motors develop very low torques and cannot be used to run high torque loads. They can, however, be used to run fans and do some useful work while helping to increase power factor. Capacitor motors are used to counterbalance inductive loads in the same manner as capacitors but are typically only useful in central locations such as the main switchgear or load centers.

Lightning and Surge Protection

Electrical surges, analogous to pressure surges in a pressurized piping system, can directly or indirectly injure personnel and often damage equipment. This occurrence of surges cannot be predicted, and special precautions must be taken to provide continuous protection for personnel and equipment. In any water utility's electrical system, different types of circuits are involved, including electrical power supply to components, low-voltage instrumentation circuits, status monitoring circuits, and signal transmission circuits. Depending on the particular use of a circuit, there is a normal voltage level at which that particular circuit and the components associated with it have been designed to operate. Electrical surges will cause short duration, high-voltage spikes on these circuits that are significantly higher than their normal or designed operating level. When surges have voltages higher than the maximum capability of a circuit's insulation, the insulation breaks down and personnel can be shocked or equipment damaged.

The three principal causes of electrical surges (called *voltage spikes* or *transients*) on electrical system circuits are atmospheric lightning, the switching of high currents within the electrical power distribution system, and electromagnetic interference (EMI). EMI is electromagnetic energy that is induced from one circuit into another. It usually consists of high-frequency AC because that is most efficiently induced by electromagnetic means.

Electrical surges are energy pulses of high voltage that last for a relatively short time—a hundredth or thousandth of a second. The higher the voltage or the longer the pulse lasts, the greater the energy content, and the greater the potential for damage. Lightning is a special case because of the enormous amount of energy. Protection against lightning requires special techniques to divert and dissipate this high amount of energy away from personnel, buildings, and equipment.

Surge protection has become more acute since the advent of telemetry and supervisory control systems. Earlier equipment designs incorporated electromechanical devices such as relays and mercury switches. Although these devices could also be affected and even damaged by the surge effects, in most cases these components had a relatively high heat dissipation capability. This meant that the heat generated during the very short period of the surge could be dissipated without permanent damage to the particular component.

Electronic equipment used in telemetry and supervisory control incorporates solid-state devices rather than electromechanical devices. The heat dissipation capability of these devices is significantly less than the electromechanical-type devices. In recent years, with the introduction of integrated circuit technology in practically every device, this problem has become more acute.

Lightning protection. Lightning is a static electricity discharge between a cloud and the earth or between two clouds. Certain atmospheric conditions cause clouds to become highly charged, either positively or negatively. When a charged cloud passes over the earth, its electrical field strength is so strong that it causes the opposite charge to build up on the earth's surface immediately under the cloud. This surface charge will accumulate in any conducting material on the earth's surface, including wet trees, wet buildings, and any metal structures, such as antenna towers or the frameworks of tall buildings. If the charge is strong enough to overcome the air's insulating ability, a lightning stroke will occur between the cloud and the charged area on the earth's surface. During the stroke, the charge on the cloud is neutralized by electricity flowing between the cloud and earth. Because the amount of charge necessary to break down the air's insulating properties is less for shorter distances, lightning tends to strike tall structures or trees.

Lightning strikes can occur at voltages of 10,000 to 500,000 volts. The stroke usually only lasts 20–500 microseconds, but the current can peak at more than 100,000 amps. Because lightning cannot be prevented, protection is provided by creating a path for these high currents to flow safely around people and equipment down into the earth.

Lightning between clouds can also cause dangerous surges in wiring systems and conducting structures. As charged clouds pass overhead, electrical charge accumulates in conducting material on the earth's surface. These charges accumulate slowly, because clouds move slowly, and the charges are *bound* by the cloud. However, if the charged cloud discharges through a lightning strike to another cloud, these bound charges in structures or trees on the earth will no longer be bound. They will, in fact, rapidly disperse, often causing currents of hundreds of amps. These high currents are known as induced strikes and can damage sensitive electronic equipment and ignite volatile fuels. Fortunately, protection against induced strikes is provided by the same systems that protect against direct strikes.

The incidence of lightning varies greatly around the country. Lightning incidence charts identify areas where lightning tends to be a significant problem. In areas of high incidence, site protection against lightning should be provided, even at remote telemetry and supervisory control installations. Reference publications are available for evaluating the desirability of site protection, such as the National Fire Protection Association handbook, NFPA 78—Lightning Protection Code. Site protection is a trade-off of cost for the protection versus consequences of the damage resulting from a lightning strike.

Lightning protection systems have three major components: air terminals, down conductors, and grounding electrodes.

Air terminals. Air terminals are short pointed rods, sometimes called lightning rods, mounted above (and sometimes on the sides) of the facility to be protected. They are designed to offer a preferred point for the lightning to strike.

Down conductors. Down conductors are designed to conduct the high lightning currents quickly and safely to the ground. They are connected to the air terminals at the top end and the grounding electrodes at the bottom end. Down conductors can be the steel framework of a structure provided that it offers a low-impedance path to the ground. Low reactance is more important than low resistance, because the current in a lightning stroke can rise as fast as 10,000 amps/microsecond. With such a fast-rising current, an otherwise trivial amount of inductance can induce a voltage drop of several thousand volts. Where adequate existing steel frameworks are not available, a minimum of two down conductors (to keep the inductance low) must be added for all structures except towers and flagpoles.

Grounding electrodes. Grounding electrodes serve to pass the current from the down conductors into the earth where the energy can be safely dissipated. These grounding electrodes function exactly like the power distribution system grounding electrodes described previously. However, because of the need to conduct such high lightning currents, the lightning grounding electrodes may be larger or more numerous.

Surge protection. While direct or induced lightning strokes cause the most powerful and dangerous electrical surges, several other phenomena can cause electrical surges large enough to damage sensitive electronic equipment. Computers and communication equipment commonly used in water utility facilities are particularly vulnerable. These other causes for surges include

- switching inductive loads, such as motors, on or off
- disconnecting high electrical currents, such as fault currents
- EMI from nearby conductors carrying large surge currents, such as lightning hitting conductors

The energy of these voltage surges is much less than that produced by lightning, and it rarely affects power distribution equipment. However, these surges often have enough energy to spark and ignite explosive atmospheres and to damage instrumentation, communication, and computer equipment.

The energy in voltage surges can directly affect sensitive equipment through the wiring systems connecting the equipment to its power supply and to long (over 100 ft) signal and communication cables. While power supply connections are an obvious path for directly applied surge voltages, long communication lines can have potentially significant surge voltages induced in them from EMI. Long cables act as antennas to high frequencies that can induce voltage surges and then conduct these surges directly into the sensitive communication equipment. Surge protection must be provided on communication, cable and signal cable terminals as well as power supply connections.

Surge energy can also be induced indirectly into sensitive equipment by EMI from the strong magnetic field produced around nearby conductors that carry surge currents. Protection against this type of EMI is typically metal encasement of the nearby conductors carrying the surge current, metal encasement of the sensitive equipment, or physical separation of several feet.

To protect against directly applied surge voltages, abnormally high voltage must be detected before it enters the sensitive equipment and routed around the sensitive equipment. Furthermore, the detection and rerouting must be done very quickly to be effective. The design of protection devices will, of course, vary widely, depending on their particular design concept. Some, of course, will be better than others in particular applications, and the user needs to determine which of the devices will be best suited for a particular application. Most of these devices are designed to be installed in conjunction with terminal strips. Surge protection devices incorporate various types of components as part of their circuitry, including spark gaps, varistors, and gas discharge tubes.

These devices are placed across the terminations of connective wiring (power, signal, and communication) and a good electrical ground. They function by shorting the connected terminals to ground whenever the applied voltage exceeds a specified safe level. The shorting of the terminals to ground allows the energy to bypass the protected equipment and dissipate harmlessly into the earth. For greater protection, the grounding conductors for surge arrestors should only be connected to other surge arrestors and not to the electrical power grounding conductors before eventually being connected to separate grounding electrodes. This will prevent backflow of the energy through power grounding conductors.

Another point that must be remembered is that all of these type of devices will require some degree of maintenance to ensure that the particular device is still active and available for protection at a specific location. Testing procedures are available, depending on the device. Also, a maintenance program should provide a regular check of surge protection equipment at defined intervals throughout the year. In addition, surge protection equipment should be inspected immediately after severe lightning storms in a particular area.

Protection against power supply surges for major equipment in central locations can be effectively accomplished through a UPS system. Although the primary purpose of a UPS is to act as a source of power during periods of loss of commercial power, a properly designed UPS system can also provide significant surge protection on the power supply (signal and communication cable terminations will still require separate surge protection) to central communications or computer systems. Again, to prevent the backflow of the surge energy, a separate isolated grounding system for this power supply is often required.

EMI signals. EMI signals are similar to surges except that they typically have a lower energy level and are continuous in time rather than occurring occasionally as with surges. These signals are continuously generated by electronic switching circuits in power systems like variable speed drives and UPS systems. EMI signals are less likely to damage equipment than voltage surges because of inherently low energy levels. However, EMI can damage very sensitive components used in low energy computer and communication equipment. Even though damage is unlikely, EMI signals can interfere with proper operation of low energy computer and instrumentation equipment; therefore, protection against this interference may be necessary. The low energy level and continuous presence of these signals preclude the use of surge protection devices and require special treatment to minimize this type of interference. General protection against it is usually effected by encasing the sensitive equipment in metal enclosures.

Because long runs of parallel wiring can cause two circuits to become electromagnetically coupled, they should be avoided. Where such runs are necessary, the wiring should be separated as far as practical and shielded, and the shields should be grounded to minimize the effect of the electromagnetic coupling.

Another common practice to protect against unusually strong signals is to provide filters at the wiring terminations of the signal-generating equipment. If that is not possible, filters should be installed at the terminals of the protected equipment. These filters are composed of capacitors and inductors arranged in a special way to block the undesirable EMI energy from entering the protected equipment or to provide a low-impedance path for the EMI energy to ground.

This brief discussion of surge protection does not provide all the answers to the design of a proper surge protection system for electrical systems. The intention of the discussion was to point out and emphasize factors that need to be considered and investigated. A number of publications that cover surge protection of instrumentation and electronic devices are available. Also, several standards that have provisions regarding surge protection, including ANSI/IEEE C37.90A and IEEE Standard 587-1980, are available. Specific problems should be referred to a competent design professional because many factors beyond the scope of this publication can significantly influence the selection of the proper lightning, surge, and EMI protection.

REFERENCES

Hopkins, M. H. Jr., and H. R. Skutt, 1972. *Introduction to Electrical Engineering*. The Ronald Press Company. Pages 1–17.

Bureau of Personnel, Department of the Navy. 1970. *Basic Electricity*. Dover Publications, Inc. Pages 1–33, 123–148, 159–249.

IEEE Standards Board. 1976. *IEEE Recommended Practice for Electric Power Distribution for Industrial Plants*. The Institute of Electrical and Electronics Engineers, Inc. Pages 25–74 and 235–251.

IEEE Standards Board. 1982. *IEEE Recommended Practice for Grounding of Industrial and Commercial Power Systems*. The Institute of Electrical and Electronics Engineers, Inc.

This page intentionally blank.

AWWA MANUAL M2

Chapter 3

Motor Controls

INTRODUCTION

This chapter introduces the basic operating principles of motors and the unique concerns associated with starting motors. Motor control involves the starting and stopping of motors and, in the case of variable speed motors, the control of their speed. The concepts of variable speed drives are also discussed. The final section of this chapter provides an in-depth discussion of motor control logic and an introduction to motor control diagrams. This basic information will enable the reader to better understand how motor-driven equipment is controlled and what operational flexibility may be available in a particular facility.

MOTORS

Motors are electromechanical devices for converting electrical energy into mechanical energy. They are usually capable of operating with alternating current (AC) or direct current (DC) electricity. However, because the vast majority of motors used in the water industry are AC, this chapter is limited to a discussion of AC motors.

Motors are usually rated in horsepower, a common unit of mechanical power. A motor's horsepower rating is the amount of mechanical power or load that can be driven by the motor without damaging it. Operating motors at power levels higher than their ratings causes them to overheat and possibly burn out. Motors can be safely operated at powers lower than their rating, but doing so lowers their power factor and efficiency.

AC motors are composed of two essential parts: stators and rotors. As the names imply, a stator is the stationary piece on the outside, while the rotor is the rotating piece in the center. Electrical energy is applied to the stator and mechanical energy is drawn off the shaft of the rotor.

Applying electrical energy to the stator causes a rotating magnetic field forcing the rotor to turn. With three-phase AC power, the stator rotates automatically as a natural consequence of three-phase power. With single-phase power, an extra mechanism must be used to cause the field to rotate. These rotation methods are described in the section *Single-Phase Motors*.

The speed with which the stator field rotates is called the synchronous speed. Synchronous speed is determined by the frequency of the applied power and the number of poles (or windings) used in the stator's construction. The synchronous speed is always indicated on the motor's nameplate.

The rotor, which sits inside this rotating magnetic field, also has an electrical current flowing through it, creating an electromagnet. The rotor's electromagnet tries to align itself with the magnetic field created by the stator, but because the stator's magnetic field is rotating, the rotor also has to rotate. The strength of the rotor and stator magnetic fields determines the torque that the motor develops and consequently the horsepower that the motor delivers.

Because these magnets are electromagnets, the strength of their fields is determined by the amount of current flowing in them. Stator currents are caused by the application of external electrical energy. The electrical current flowing in the rotor is produced in one of two ways: by induction from the stator's magnetic field or by connecting an external source of electrical energy to the rotor. Each of these methods operates a little differently and gives rise to two basic kinds of AC motors: the induction motor and the synchronous motor.

Induction Motors

Induction motors are made in two basic varieties that are based on the construction of the rotor. They are called the squirrel-cage induction motor and wound-rotor induction motor. The squirrel-cage induction motor has its name because the rotor resembles a squirrel cage. It is made out of two metal rings connected by several straight parallel metal bars (see Figure 3-1). The wound-rotor induction motor (also shown in Figure 3-1) has windings made of wire on the rotor which are similar in construction to the windings in the stator. This type of rotor often has external connections to allow the electrical current in the rotor windings to be controlled. This permits speed control and, in some cases, energy recovery. The external connections are made through rings mounted at the end of the rotor's shaft. Sitting on each ring is a stationary metal piece called a brush. The brush touches the ring and makes an electrical connection while allowing the ring to slip under the brush as the rotor rotates. Consequently, the rings are often referred to as slip rings. The stationary brushes can then be connected by wires to control circuits. Because the slip rings are always moving while contacting the brushes, the brushes wear out and must be replaced periodically.

Induction motors are self-starting. As soon the stator voltage is applied, the rotating field begins to move through or *cut* through the rotor conductors. This movement causes very high electrical currents to flow in the rotor conductors, which creates a very strong magnetic field of its own. These induced currents and the resulting magnetic field create a force that causes the rotor to try to turn in the same direction as the stator field is rotating. As long as the force is strong enough to turn the rotor shaft and the load connected to it, the motor will start. As the motor increases speed, the relative speed between the stator field and the rotor will decrease. This lower relative speed will cause the rotor current to decrease, resulting in a decrease of the magnetic field strength of the rotor. This decrease reduces the force (called *torque*) causing the rotor to turn. The rotor will increase speed until the torque reduces to the amount required to turn the load. However, if the rotor turns as fast as the stator field (the synchronous speed), the stator field will stop moving through the rotor conductors; their relative motion will be zero. No current will flow in the rotor conductors, and no torque will be produced to turn the load. An induction motor cannot rotate at synchronous speed and must operate at less than synchronous

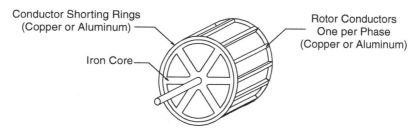

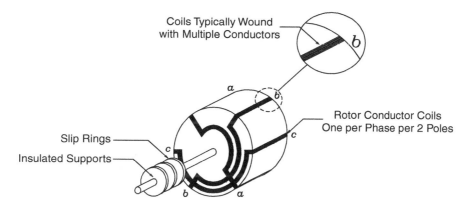

Figure 3-1 Induction motor rotors

speed. How much less is referred to as *slip*. Slip is expressed as a percent of synchronous speed and is usually abbreviated by a small *s*. The equation for slip is

$$s = \frac{(\text{synchronous speed}) - (\text{running speed})}{(\text{synchronous speed})}$$

Because the amount of slip is determined by the amount of torque necessary to drive the load, the speed of an induction motor varies as the load varies. Because slip values are usually below 5 percent, the speed variations are small.

The wound-rotor induction motor has several disadvantages compared to the simpler squirrel-cage induction motor. The wound-rotor induction motor is more expensive to build, heavier, and has the added maintenance requirement of brush replacement. The squirrel-cage induction motor is usually preferred. Its maintenance simplicity, low initial cost, and versatility of mounting make it the best choice in most installations. In older installations where large horsepower sizes (greater than 500 hp) and variable speed control or energy recovery were necessary for economic or process reasons, the lower-efficiency wound-rotor induction motor was generally used. However, recent advances in the development of higher-power semiconductors have improved the power-handling capabilities of variable frequency controllers. This improvement allows for the use of the more efficient induction motor in variable speed applications requiring the higher horsepowers.

Synchronous Motors

Synchronous motors are essentially AC generators connected backwards and constructed similarly as a wound-rotor induction motor. However, synchronous motors are more difficult to operate than induction motors. They can only run at synchronous speed (a slip of 0 percent) and require separate DC excitation of their rotor windings.

Synchronous motors require special starting equipment for them to reach synchronous speed before they will run on their own. Sometimes a small separate motor is used to bring the motor up to speed, but it is more common to build a small induction motor into the synchronous motor that acts as a starting motor and is then disconnected when the motor reaches synchronous speed.

Synchronous motors must have a DC power source to make its rotor an electromagnet. A DC generator is usually mounted on the shaft of the synchronous motor. The generator's output is connected to the rotor windings through a control circuit that allows the DC rotor current to be adjusted.

Synchronous motors are complicated and expensive compared with induction motors. However, synchronous motors are used for two reasons. First, synchronous motors only run at synchronous speed. If precise speed control is required (as with drum programmers, clocks, or tape players), the synchronous motor does a better job than an induction motor whose speed is affected by the load. This reason accounts for the use of most small synchronous motors. Second, and of much greater importance in industrial applications, synchronous motors can be operated with a leading power factor. This means they can be used to offset the lagging power factor of induction motors. Therefore, if a facility uses mostly induction motors for all but the largest motors, then one or two large synchronous motors can be used to bring the power factor closer to one. This arrangement can save considerable energy costs.

Single-Phase Motors

Single-phase motors are almost always induction motors, but they require extra mechanisms to cause the stator field to start rotating. When the rotor begins to turn, the rotor field interacts with the stator field to keep the field rotating. Consequently, it is only necessary to get the motor started. Because starting mechanisms use a little energy, they are usually disconnected after starting, so the motor can run more efficiently. The three most common types of single-phase motors in use are explained below.

Split-phase motors. These motors use an extra stator winding that is installed in a way that causes the starter winding field to interact with the main stator winding's field resulting in a small strength rotating field. This causes the rotor to turn. When it is near synchronous speed, a centrifugally operated switch disconnects the starter winding. These motors have moderate starting torques and good speed regulation over their designed load range. Consequently, they are widely used in small electrical equipment that do not require high starting torques.

Capacitor motors. These motors are made in the same way as split-phase motors except they have a capacitor installed in series with the starter winding. This capacitor in series causes a stronger rotating field and, therefore, provides a greater starting torque than the split-phase motor. Capacitor motors are made both with and without the centrifugal switch to stop the starter winding after the motor is up to speed.

When the switch is used, the motor is called a *capacitor start* motor, and when the switch is omitted leaving the starter winding energized, the motor is called a *capacitor run* motor. Omitting the switch saves a little cost and increases reliability although there is a small reduction in motor efficiency. Capacitor motor speed is more sensitive to load variations than split-phase motor speed, but the capacitor motor's higher starting torques make them the preferred choice for use on small electrical equipment that require high starting torques like compressors, grinders, drill presses, and small conveyer belts.

Shaded pole motors. These motors use a small conducting ring inserted in the stator's pole pieces. A current is induced in the ring creating a small magnetic pole in the shadow of the main pole's magnetic field. This *shaded* pole's magnetic field interacts with the magnetic field of the main pole. The effect of this shaded pole is not to rotate the field around the rotor, but to cause a small rotating field around the pole piece. This field rotates at synchronous speed and induces a small current in the rotor causing it to rotate. This design has the advantage of being very inexpensive to make but produces the lowest starting torque, has the greatest speed sensitivity to load variations, and has the lowest efficiency of any single-phase AC motor. Consequently, this type of motor is only made in very low horsepowers (less than $1/20$ hp) and is primarily used to drive equipment fans.

Motor Starting

Starting a motor is always more difficult than keeping it running or stopping it. Consequently, some special techniques are used to start motors. To understand what is required to start a motor, the starting torque, starting current, and starting voltage must be considered. Each of these aspects of motor starting is described in the following sections.

Starting torque. Starting torque is the mechanical force necessary to move the load connected to the motor. The heavier the load, the more difficult the motor is to start. Motors are designed to develop more torque when they are started than when they are running. How much torque is developed as the motor goes from zero to full speed depends on the motor's design.

The National Electrical Manufacturers Association (NEMA) has defined four standard starting torque curves for motors. These are referred to as the NEMA type A, B, C, or D torque curves. Most motors are designed to produce one of these four torque curves, and each type is designed to start loads having different mechanical characteristics. This information is always stamped on each motor's nameplate.

Starting current. Torque developed by a motor is directly proportional to the current flowing in the motor's rotor and stator. Therefore, to develop a higher torque when starting, a motor must draw a higher starting current than after it's running. Starting current always starts out very high when the motor is first connected to a voltage. Then, as it begins to rotate, the current drops rapidly until it stabilizes at the running value.

Motor nameplates always show the current a motor draws while running at its rated horsepower. This current is called full load amperage (FLA). The actual starting current is several times this value and varies with each motor and load combination. However, a good rule of thumb for induction motors is that the starting current is initially six times FLA and then declines to the running value as the motor reaches running speed.

Starting voltage. Motors are always designed to be connected to a standard value of distribution voltage. This value is always stamped on the motor's nameplate. Common values for three-phase motors are 208, 230, and 460 volts. This is the value of the voltage that should be available at the motor's power feed terminals when it is running. Because the motor current is delivered to the motor through wires from the actual source of electricity to the motor, some voltage drop will occur on those feeder wires. The amount of the voltage drop will depend on the motor's FLA and the amount of resistance in the wires feeding the motor's power. Because a motor's starting current is several times more than the running current, the amount of voltage that is dropped on the feed wires during starting will be several times greater than when it is running. These wires need to be the appropriate size to ensure that there will be enough voltage at the motor terminals to start it.

Some motors have dual voltage ratings such as 230/460 volts. These motors will also have dual FLA ratings such as 28/14 amps. This means that the motor will draw more current if operated at the lower voltage and less current if operated at the higher voltage. However, in both cases the motor will develop the same horsepower.

Motor Starters

Motor starters are specially designed switches used to start motors. Motor starters can be mounted with a group of starters in a large enclosure called a motor control center, or they can be mounted individually in wall- or floor-mounted enclosures. They are designed to safely switch the necessary power to the motor and to prevent the motor from drawing too much current.

Motor overload protection is necessary because if a motor draws too much current, it will become excessively hot. This extra heat will reduce the motor's life and may cause it to burn up quite rapidly and possibly start a fire. Several things can cause a motor to draw excessive current. If the motor is restricted from turning at near synchronous speed, it will draw too much current. This might be caused by a mechanical problem with the load or failure within the motor itself. In any case, the motor's current must be limited to those values which the motor can safely withstand. Therefore, all motors must be provided with some type of current-limiting device to prevent the motor from overheating.

Ideally, it is desirable to prevent a motor from drawing more than the FLA stated on the motor's nameplate. However, the high starting current must also be allowed for. To provide protection against a small overload current and to permit the high starting current to flow, a timed overload concept is used. Timed overload means using devices that can sustain a lower overload current for a longer time than a high overload current. Timed overload permits the starting current for a short time (a few seconds) but will stop the motor if the overload current persists for too long. These devices will also permit a small overload current for several minutes but will eventually stop the motor if the small overload current persists for too long.

The devices that provide this motor protection are called *overloads* because they protect the motor from being overloaded. The motor's feed current is usually passed through bimetal strips. When too much current flows, the strips heat up, causing them to bend. When they bend too much, they activate or trip a small switch disconnecting the power from the motor. These devices are commonly called *heaters* or *motor trips*. Whenever these devices trip a motor, they latch and hold the motor off. They must be reset, similar to a circuit breaker, to restart the motor. A reset button is usually found on the front of the motor starter.

Motor starters are classified by what operates them, how big a motor they can safely switch power to, and by whether or not they also include the branch circuit disconnect and protection device. These categories are discussed in the following sections.

Motor starter operation mechanisms. Because motor starters are primarily switches, it is useful to know what causes the switch to activate. Motor starters are made in two primary types—hand operated or electrically operated. Hand-operated motor starters usually have a lever on the side of the starter to turn it on or off. Bigger starters must have bigger levers and, sometimes, windup springs to make it easier to move the larger mechanisms. Hand-operated motor starters are usually only used on smaller motors because they cannot be operated remotely, and they will cause the motor to immediately restart after a power failure. Instantly restarting several motors after a power failure can be dangerous because of the high starting currents used to start all of the motors at once. For these reasons, most motors starters have electrically operated switches.

An electrically operated switch is actually an electromechanical relay. Relays used to switch power-carrying conductors are called *contactors*. Contactors contain a coil, which, when energized by electricity, becomes an electromagnet that pulls the switch contacts closed. The coil is de-energized when the contacts are pulled apart by springs. The motor can be started or stopped by anything that can energize or de-energize the starter's coil. When this is done manually by using a small switch or button, the effect is the same as with the hand-operated motor starter. The advantage is that the switch or button can be located in a more convenient place than the starter and connected to the starter by wires. This type of starter can also be operated in a way so that the motor will not restart after a power failure until the button is pushed again.

Motor starter sizes. Motor starters have to be able to switch the necessary power to the motor. Because motors come in many sizes, so do motor starters. NEMA has developed standardized motor starter sizes and a group of horsepower ranges called by numbers. The smallest starter is *00* (double ought) for fractional horsepower motors. The next larger size is called *0* (zero or one-ought), and the next larger is called *1,* and so on. The largest standard size is called *9* which can start a 1,600-horsepower motor. Larger motor starters are also made, but they are outside the standard NEMA sizes.

Motor feeder protection. Motor current is carried to the motor from the starter by motor feeder wires. The electrical circuit made up of the motor starter, the feeder wires, and the motor are called a *branch circuit*. Branch circuits are required to have overcurrent protection and a disconnecting means according to the National Electric Code (NEC; see chapter 2). When this branch circuit protection is included with the switching mechanism and overload protection in a motor starter, the starter is called a *combination starter*.

Motor Starting Techniques

Different types of motors require different techniques to get them started. A general discussion of the special requirements for synchronous motors and single-phase motors was described above. However, the vast majority of motors used in a water utility's facilities are of the three-phase induction type, and some of the different techniques used to start three-phase induction motors are discussed below.

Full voltage (across the line). Because three-phase induction motors are inherently self-starting, all that is necessary to start them is to apply the operating voltage to the motor terminals. This full voltage starting is sometimes referred to as *direct on-line* (DOL) or *across-the-line* starting. DOL is by far the most common way to start three-phase induction motors.

Reduced voltage. With large motors (over 100 hp), the facility's electrical power distribution system or the motor's feeder wires may be too big to allow for the high starting current. In some cases, the electric utility has regulations preventing a motor that is too large from being started across the line; otherwise, a voltage drop might occur in the electric utility's distribution system when the motor is started.

Reduced voltage starting is used when a voltage drop may occur. Reduced voltage starting temporarily reduces the starting voltage until the motor begins to rotate, and then the applied voltage is switched back to its regular value. This procedure limits the starting voltage and the starting current so that bigger wires will not have to be used. Reduced starting voltage also reduces starting torque, which must be considered when using this method. The switching action required to support reduced voltage starting takes place automatically in specially designed motor starters called *reduced voltage starters*.

Several different types of reduced voltage starters are commonly used including autotransformer, primary resistance, partial winding, and wye delta. Each type provides a different treatment of the starting voltage and current and the method of acceleration to accommodate different driven load requirements.

Bidirectional. Sometimes process requirements require the ability to operate a motor in two directions. This bidirectional ability is commonly done in valve actuators, where one direction closes the valve and the other direction opens the valve. All that is required to change the direction of a three-phase induction motor is to switch any two of its three feeder wires. Therefore, starters designed to start a motor in either direction simply have two contactors with two wires crossed on one of them. As an added safety measure, they also have a mechanism for preventing both contactors from being energized at the same time. These starters are called *bidirectional* and are made in both full voltage or reduced voltage versions.

Multispeed. Some process conditions require the load to run at two or three different speeds. Multispeed motors are often useful in pumping systems where two different rates of pumping can best serve the range of operation. Motors can be made with two or three sets of stator windings around a single rotor. Each winding is arranged to cause a different synchronous speed and, therefore, a different running speed. This permits the same pump to be operated at different speeds and, therefore, different pumping capacities. To start these multispeed motors, a separate contactor must be used for each set of windings. The two-speed starter has the same components as a bidirectional starter but is wired differently. The three-speed versions have three contactors in them. As with bidirectional starters, these are made in both full voltage and reduced voltage versions.

Insulation Type and Service Factor

When motors are running, they get hot. They are designed to withstand a specific amount of heat without damage. The actual temperature of a running motor depends on how heavily it is loaded, how often it is started, how long it has been running, and how hot the environment is where the motor is located. This means even if they are not overloaded, they may still get too hot. The primary concern with too much heat is the breakdown of the insulation, which insulates the wire in the motor's windings. Two factors in a motor's design and manufacturing determine the amount of heat the motor can safely withstand—insulation type and service factor.

Insulation type. The insulation used has a specific ability to withstand a certain amount of heat. Insulating material that can withstand higher temperatures is more expensive, so motors are made with different insulating materials. Each type of material has a different cost and temperature withstand rating. The temperature ratings have been standardized by NEMA and are identified by letters. The insulation type used is always identified on the motor's nameplate. Motors that operate in high ambient temperatures, such as outdoors in hot climates, will need to have the more expensive insulation.

Service factor. Motors that run all the time never get a chance to cool down; motors that run only occasionally never run long enough to get hot. A good example of the latter is a valve or gate operator. Therefore, depending on the service, it is possible to safely use motors that do not have a high temperature rating in applications that would otherwise require it.

The measure of how much time the motor can be operated safely is called the *service factor*. A service factor of 1.00 represents all the time and is sometimes called *continuous duty*. Service factors less than 1.00 mean the motor is not designed to be run continuously. If motors with a service factor less than 1.00 are run continuously for long periods, they may burn up even though they are not overloaded. Conversely, motors with a service factor greater than 1.00 have an extra ability to withstand heat. This is useful for motors expected to be started frequently and run most of the time, which actually makes them hotter than if they were to run continuously. The extra heat is caused by the high starting current. It does not damage motors to be run at a service factor less than they are rated for, but it wastes the extra cost associated with buying the higher service factor motor.

Motor Disconnects

Motor disconnects are an important safety requirement and are required by the NEC. If maintenance is performed on a motor-driven piece of equipment, there should be an absolutely reliable method in place ensuring that the motor cannot be turned on while the maintenance is performed. Motor disconnects are used to accomplish this. They are switches that break the path of each motor feed wire. These switches must be located such that no other source of electricity can bypass the switch and energize the motor.

The NEC requires the motor disconnects to be within sight of the motor. "In sight of" is defined as having a clear and unobstructed view from the motor's location. The NEC further limits the distance of the disconnects from the motor to 50 feet. A disconnect switch can be located more than 50 feet from the motor or out of sight from the motor *if* the disconnect switch can be padlocked in the OFF position. Maintenance personnel can then use their own padlock to lock the switch in the OFF position guaranteeing safety.

Motor control circuits sometimes use a lock-out stop switch located at the motor. This switch is often abbreviated LOS. An LOS cannot prevent the motor from being energized in all circumstances. Therefore, *the LOS is not a motor disconnect switch and must never be used to lock a motor for maintenance purposes*. These LOSs are provided only as an operational convenience to prevent equipment from being operated if there is a reason not to use it.

VARIABLE SPEED MOTOR CONTROL

Variable speed drives are the general class of equipment used to drive mechanical loads at varying operational speeds. This equipment must be controlled by a signal based on a process parameter such as level, pressure, or flow, as detected by one of the methods discussed in other chapters of this manual. The signal from these measurement systems is used to control the actual load speed by controlling a variable speed drive that controls the load's speed. Variable speed drives fall into two general categories: variable torque transmission systems and variable speed motor control systems.

Variable Torque Transmission Systems

Variable torque transmissions are used to mechanically couple fixed-speed motors to mechanical loads. Consequently, they have fixed-speed input shafts with variable speed output shafts. A distinct disadvantage of these systems is their need to dissipate energy in the form of heat whenever the load is operated at less than full speed. Consequently, they have very poor efficiencies when working over their operating range. Speed range is approximately 50 to 95 percent of the rated motor speed. The motor should be selected as close as possible to and above the maximum of the desired speed of the load. Using a motor with a rated speed much higher than the load requires will result in high slip, with a significant amount of input horsepower to the motor resulting as heat rather than passing to the load.

These mechanisms compared favorably with wound-rotor speed control technology. However, the development of solid-state semiconductors has made variable frequency drives, which have a higher efficiency, the preferred choice for small- to medium-size motors (less than 1,000 hp). Another advantage of torque transmission systems is that they can be manufactured to accommodate very high horsepowers. The two most common types in use today are eddy current clutches and liquid clutches. Other types of torque transmission equipment have been used occasionally, including clutches using water as the hydraulic fluid and hydraulic pump-motor combinations. Most have proven unsatisfactory or have no particular advantage over the more commonly used devices.

Eddy current clutches. The eddy current clutch consists of a constant speed input shaft directly close-coupled to a synchronous or induction motor and a variable speed output shaft connected to the load. An electromagnetic member is connected to the output shaft, and a field coil member is connected to the input shaft. The field coil member surrounds the electromagnetic member. As increased speed is required of the load, DC current is introduced into the electromagnet, exciting it and causing eddy currents. The magnetic flux produced by the magnet develops torque, tending to turn it and the connected load in the same direction as the developed torque, therefore producing an output speed proportional to the amount of DC current. The DC power applied to the field is controlled by a current signal. This signal may come from any measurable variable. Clutch efficiency is 2 to 3 percent below wound-rotor motors with secondary resistance.

Eddy current clutches are available in a full range of sizes and speeds in both horizontal and vertical configurations. They provide reliable speed control for remote manual or automatic control. However, this type of drive is not encouraged for applications where the motor drive is mounted above the driven load and directly to it; for example, directly to the pump frame of vertical high lift pumps. The weight of the combined unit (motor and clutch) is frequently greater than that of the pump, causing imbalance and requiring special bracing.

Liquid clutches. Liquid clutches are called *hydroviscous drives*, essentially a wet clutch that transmits energy through the shear strength of oil. These units are comparable to the eddy current clutch with slightly higher efficiency at full speed and a lesser efficiency at reduced speeds and can be manufactured with very large horsepower ratings. The primary use of these units has been for large plant pumping systems.

VARIABLE SPEED MOTOR CONTROL SYSTEMS

Variable speed drives allow the motor to be close-coupled to the load because they cause the motor to vary in speed. The two most frequently used techniques are variable frequency controllers with induction or synchronous motors and secondary energy recovery units with wound-rotor induction motors. These systems are becoming the most popular choices for variable speed drives because they have efficiencies in excess of 85 percent throughout their operating ranges. These high efficiencies are possible because these systems do not have to waste energy in the form of heat as with variable torque transmission systems and outdated non-regenerative wound-rotor motor controllers. However, these systems rely on solid-state power semiconductors and are limited in their voltage-handling capabilities. Motors rated above 1,000 hp typically require the use of voltages in excess of 600 VAC. These higher voltages (e.g. 2,300 and 4,160 VAC) require very expensive and very large electronic equipment. These size and initial cost factors must be weighed against the expected operational savings from operational efficiency to determine if they are economically justifiable.

Wound-rotor motor controls. When wound-rotor motors are used, control is accomplished by controlling voltage on either stator or rotor windings. Control of voltage on rotor windings is preferable on motors over 75 kW (100 hp) or on voltages over 480 to reduce the amount of power the control must manage. Depending on the control system, a restricted full speed may or may not be experienced with these motors.

The advantages of these types of drives are that the current draw during starting is greatly reduced, and the mounted weight of the motors, while generally slightly greater than that of a normal squirrel-cage induction motor, is considerably less than that of a motor–clutch combination.

The primary method used to vary the rotor voltage is to vary the impedance of the rotor windings. Older systems used techniques such as resistors, reactances, liquid rheostats, and more recently, solid-state electronic rheostats. However, all of these methods required the dissipation of heat whenever the motor was run at less than full speed. Consequently, the efficiencies of these older designs were quite low.

The development of the solid-state electronic rheostat led to regenerative controllers. With regenerative controllers, most of the energy that was wasted as heat is now converted to fixed frequency AC power and fed back into the plant's power grid. The average efficiency of this method is usually between 85 and 90 percent. Because this method applies solid-state technology only to the energy drawn off the rotor at operational speeds below 100 percent, it can handle larger motors and higher excitation voltages with smaller power semiconductors than variable frequency controllers.

Variable frequency controllers. Variable frequency controllers are true variable speed drives in that the motor will vary its output speed with or without the load connected. In operation, these supply systems use power semiconductors 1) to change three-phase, 60-Hz power to DC (rectification), filter the DC, and 2) to restructure (invert) the DC into an output of variable frequency and voltage.

Two basic types of variable frequency controllers are available: *current sourced* and *voltage sourced*. These terms refer to the technology used in the DC and inverter sections of the controller. Current-source controllers are smaller and cheaper than voltage-source controllers but must match the reactance of the motor's windings. Furthermore, current-source controllers cannot control more than one motor at a time. Because of these limitations, voltage-source controllers are preferred in sizes below 500 hp where their size and cost premiums are offset by the easier coordination of manufacturing and flexibility of application. Voltage-source units are not available for motors operating above 600 VAC.

An important consideration in the use of these controllers is the shape of the output waveform. The shape is not a smooth sine wave like that produced by a rotating generator. The wave is produced by switching DC back and forth through the motor's windings, producing considerable energy in the harmonics of the output frequency.

Harmonics are higher frequencies that are an even multiple of the base frequency. If the base frequency is 60 Hz, the harmonics will be 120 Hz, 180 Hz,

240 Hz, etc. Harmonics are not produced when rotating generators are used, but they are always produced whenever electricity is switched.

A number of techniques are used to minimize harmonic output and improve power factor and efficiency. Some units have high efficiency but have relatively poor power factor. Others have a constant high power factor but a reduced efficiency. Some systems require motors with additional thermal capacity because motor heat caused by excessive harmonics will occur. Even with good harmonic filtering the motor should be rated for a 1.15 service factor and be provided with high-temperature insulation.

This type of drive permits the use of standard squirrel-cage induction motors, allowing application in virtually any location or in any mounting configuration. Furthermore, it will draw power only as necessary to drive the output at the associated speed with no heat generation caused by power wastage as found with eddy current or hydraulic drives. The efficiency of the controller over normal ranges will generally stay above 92 percent. The loads can be operated at any speed up to 150 percent of the standard motor's full load speed. However, operation above 100 percent of the motor's rated speed (associated with 60 Hz) will cause the motor to overheat and is not recommended as a normal operating strategy.

To prevent harmonic feedback problems associated with variable frequency drive equipment, variable frequency drive control units should always have isolation transformers on their input circuits to isolate them from the plant's electrical system. In some cases, filters are also required. In addition, each drive should be on a separate circuit. The feedback can affect electrical power bus protection relays, computerized control and data acquisition hardware, and radio communication. Also, generator manufacturers recommend derating generators when a substantial portion of the electrical load supplied by the generators is for variable frequency drive equipment.

MOTOR CONTROL LOGIC

To turn motors on or off, the electricity must to be connected to or disconnected from the motor. As previously discussed, the actual switching of the power to the motor windings is performed by the starter. Any special starting requirements of the type of motor are also handled by the motor starter. The techniques or schemes used to cause the starter to start or stop the motor and why they are used is discussed in the following paragraphs.

Because many different schemes, each dependent on the application, are used to control motors, a special kind of diagram has been developed to document the particular method used in each case. These are called motor control diagrams. As each basic type of control scheme is discussed, a diagram will be shown. This method will show not only what is commonly done and why, but also how to read and interpret the control diagrams for the various control schemes used in water facilities.

As stated previously, a motor will start and run whenever the starter contactor's coil is energized. Conversely, the motor will stop running whenever the starter contactor's coil is de-energized. Therefore, to control a motor, its starter contactor must be controlled.

Each electric wire is shown diagrammatically by using a line in a motor control diagram. Each path through which current can flow is shown as a separate line. All relay coils are drawn by using small circles. When the relay in question is a motor starter contactor, the circle is labeled with an *M* for motor as shown in Figure 3-2.

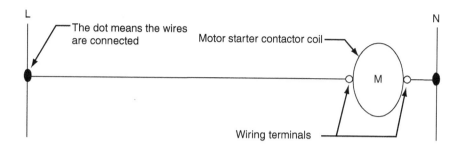

Figure 3-2   Motor starter contactor coil

Figure 3-3   Motor starter circuit with one switch

The *L* means line or source of electricity. The *N* means neutral or return path to the source. The energy used to power motor control circuits is very small compared to the power required to run the motor. Therefore, the control power voltage is usually only 120 VAC or 24 VDC. This control power can come from a separate circuit breaker in a panel, but this power is usually drawn from the starter through a small transformer called a control power transformer.

In the above example, the motor will run continuously because the current can flow through the contactor coil whenever the power (or line) is turned on. No switch controls whether the contactor's coil is energized. To correct this obvious oversight, a switch can be put on the *line* side of the contactor's coil as shown in Figure 3-3. *The neutral is never switched—only the line can be switched.* This is because the neutral is also grounded in modern systems for safety reasons, and therefore, the ground could carry the current even if the neutral was switched open.

Now S1 (switch number 1) can connect the line (turn on) or disconnect the line (turn off) from the contactor's coil. The same intuitively obvious system is used to illustrate how a motor runs in more complicated diagrams. That is, anything that can cause a complete path from the line side of the diagram to the contactor's coil on the diagram will cause the motor to run. The motor in this example (Figure 3-3) is therefore controlled by S1, called two-wire control because S1 has two wires.

With a switch at the motor and on a control panel as shown in Figure 3-4, then, if either S1 or S2 were closed, the motor would run. But notice that if S1 is closed, it does not make any difference whether S2 is closed or not; the coil will still be energized. Therefore, *both* switches must be open to stop the motor. This switch is shown in Figure 3-4.

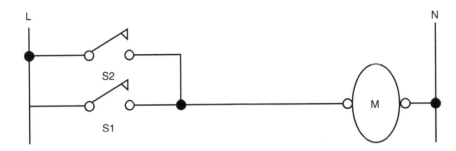

Figure 3-4 Motor starter circuit with two switches

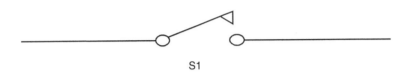

Figure 3-5 Maintained contact switch symbol

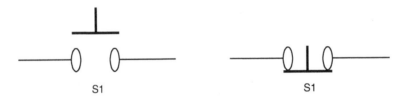

Figure 3-6 Momentary contact switch symbols

The symbol in Figure 3-5 represents a maintained contact switch. That is, if closed, then it stays closed until opened. Conversely, if opened, then it stays open until closed. A home light switch is a common example of this kind of switch.

Another kind of switch is called a *momentary* or *spring-loaded switch*. This kind of switch will only stay in one of its positions, not either position, when it is not being touched. Refer to Figure 3-6. The switch on the left is always open unless pushed down against a spring. When pressed, it closes; when released, it opens. This switch is called *normally open*. The one on the right is *normally closed,* and it opens only when pressed against a spring.

If a normally open momentary switch is used to start the motor as shown in Figure 3-7, the motor will only run while S1 is pressed. Obviously, this switch is not too useful if the motor needs to run more than a few seconds. Therefore, another mechanism is needed to keep the motor running.

A device called a *control relay* is used to solve this problem. A relay is a coil of wire that creates an electromagnet. When electricity is connected to it, the coil is energized or *picked up*. When a relay picks up, it changes the position of one or more switches, which are connected to it through levers. A control relay's coil is shown diagrammatically as a circle just like the motor contactor relay. It is labeled CR, which stands for control relay. If there is more than one, then they are numbered (i.e., CR1, CR2, CR3) as shown in Figure 3-8.

Figure 3-7 Momentary start switch circuit

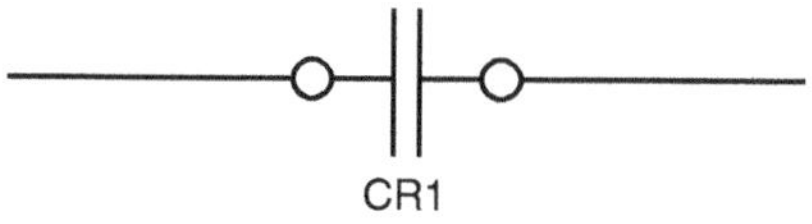

Figure 3-8 Control relay coil symbol

Figure 3-9 Control relay contact symbols

Switches operated by relays instead of people are called contacts and are available in two basic forms, as shown in Figure 3-9. The CR1 and CR2 identify the relay coil which operates them. The one on the left is the symbol for a normally open contact. This means if the relay is not energized (i.e., de-energized), then the contact is open. When the relay picks up, it closes the contact and holds it closed until the electricity is disconnected from the relay's coil. The contact on the right is called normally closed and it operates in an opposite manner. That is, the contact is closed until the relay picks up, and it is open until the relay's coil is de-energized. Note that the normal state of a relay contact is the state that the contact is in when the relay's coil is de-energized.

Figure 3-10 shows a sample use for a control relay.

As drawn, the motor cannot run unless CR1 is picked up. The only way to pick up CR1 is to move the electricity to the left side of CR1's coil. The only path that can do that is through S1. If S1 is pushed to close it, CR1 will pick up and close *both* of its normally open contacts.

The lower CR1 contact causes the motor to run and the CR1 coil will be able to stay energized because the upper CR1 contact provides an alternate path around S1 for electricity to move to the left side of the CR1 coil. If S1 is released and it opens up, the motor stays running because the upper CR1 contact allows the current to continue to flow around S1 and keep the CR1 coil energized. When a relay can energize itself through one of its own contacts, it is *electrically latched*. The contact used for this purpose is often called a *holding contact* because it holds the relay in an energized state.

To stop the motor, the CR1 coil must de-energized. The electricity energizing CR1 must pass through S2, which is a normally closed momentary switch. If S2 is

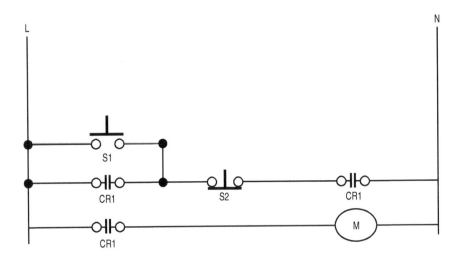

Figure 3-10 Three-wire motor control circuit

pushed for one second, then CR1 will drop out or de-energize. This opens both of the CR1 contacts, which in turn stops the motor and removes the path around S1. Therefore, when S2 is released and it recloses, CR1 stays de-energized, and the motor stays off.

By alternately pressing S1 and S2, the motor can be started and stopped. This kind of control is called three-wire control because the two on–off switches have three wires connected to them.

Control relays are more complicated than the two-wire circuit but are useful in certain situations; for example, when the power fails and comes back on. In two-wire control circuits, the switches have maintained contacts so the motor restarts when the power comes back on. If several motors were running when the power failed, they will all start as soon as the power comes back on. If there were several small motors or a few large motors running, the collective starting current of all the motors may overload the electrical distribution system and cause it to shut down. If three-wire control was used, the S1 switch would have to be pushed for each motor to restart it. Because they all cannot be pushed at once, the starting currents will not happen at the same time. Consequently, the electrical distribution system will not be overloaded.

Another reason for using control relays lies in the second two-wire control example shown in Figure 3-4. Here, the idea was to start (or stop) the motor from two different places, perhaps at the motor and at a control panel. The problem is the motor must be stopped at the same place it was started, a nuisance if the motor is a long way from the control panel.

A three-wire circuit with two locations for control should be considered, as shown in Figure 3-11. Either S1 or S3 can pick up CR1, latch it, and start the motor. Then either S2 or S4 can drop out CR1 and stop the motor. In fact, any number of switches could start and stop the motor, and they would all work independently of each other.

Notice how, in each diagram, the source of power (the line) is always drawn down the left-hand side of the diagram, and the return path (the neutral) is always drawn down the right side. These are thought of as the sides of a ladder with each path for electricity drawn across the diagram, like a rung on the ladder. This kind of diagram is called a *ladder diagram*.

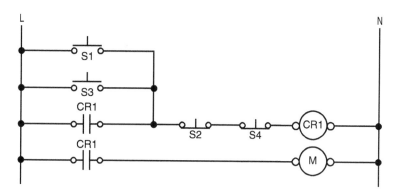

Figure 3-11 Three-wire motor control circuit with two control locations

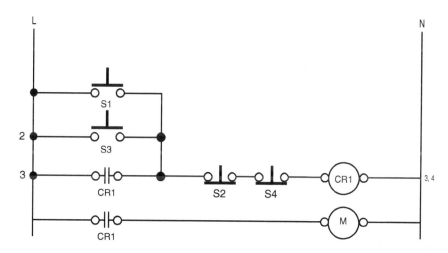

Figure 3-12 Ladder diagram with line numbers

In the examples given so far, there have been only a few rungs and only one relay. However, many motor control circuits are far more complex. Therefore, the rungs are usually numbered from top to bottom on the left side. On the right side of each rung containing a relay, the number of each rung which has a contact driven by that relay is given. If the contact is normally closed, the rung number is underlined, and if it is normally open, the rung number is not underlined.

A two-location, three-wire control circuit would look like Figure 3-12.

Status Indicators

An indication of whether or not the motor is running is important. The current status of a motor, running or stopped, is particularly useful if a start switch is located remotely from the motor, such as on a motor control center or on a main control panel somewhere. The most common way to indicate a motor's status is to use a small colored light. Usually a red light is used to indicate the motor is running and a green light is used to indicate that the motor is stopped. Status indicating lights are shown diagrammatically in Figure 3-13.

The letter in the circle indicates the color of the light, such as R = red, G = green, A = amber, and W = white. The running light can simply be wired across the

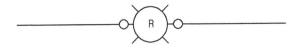

Figure 3-13 Status indicating light symbol

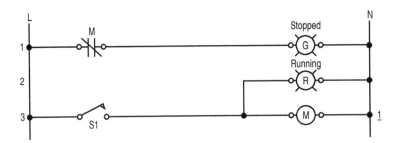

Figure 3-14 Motor circuit with indicating lights

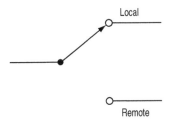

Figure 3-15 Selector switch symbol

starter contactor's coil so if the starter coil is energized, the running light is also. The stopped or off light can be energized through a small auxiliary normally closed contact of the motor contactor. In the first two-wire control example used above, running and stopped indicating lights are shown (Figure 3-14).

If additional indicators are needed at other locations, they can simply be wired in parallel with the ones shown in Figure 3-14.

Local–Remote

Figure 3-14 shows how to use more than one switch to control a motor at the same time. However, sometimes limiting the operation of a motor to one location or another is desirable. For example, it may be desirable to have a switch at the motor to use if an operator or a maintenance person is in the field, but most of the time the motor is controlled from a remote control panel. To control where the motor is operated from, a selector switch is used that has two positions—local and remote. A selector switch is shown diagrammatically in Figure 3-15.

These switches are maintained contact switches. In fact, whenever a selector switch symbol uses a small arrow, this indicates that the switch is a maintained contact switch. This means the switch stays in whatever position it was set, and it can only be in one position at a time. Selector switches may have any number of

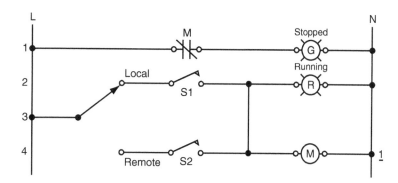

Figure 3-16 Motor circuit with local–remote switch

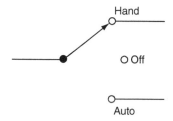

Figure 3-17 Hand-off–auto switch

positions, and the positions can have different names. The name of each position is written on the diagram similar to the local–remote example shown in Figure 3-15.

When a local–remote switch is used, it is located next to one of the operating switches. This switch is then considered the local switch and the other switch is considered to be the remote switch. They could be wired as shown in Figure 3-16.

This switch allows only one of the two switches, S1 or S2, to operate the motor. Which one operates the motor depends on which position the switch is left in. The local position allows only S1 to be used to control the motor while the remote position allows only S2 to control the motor.

Automatic Control

What has been described so far are some simple techniques for starting and stopping motors manually. Starting or stopping some motors automatically is useful. The ability to control the motor manually is usually retained. That is, the motor control is provided with a selector switch to select either manual or automatic control. A very common selector switch used for this purpose is called a *hand-off–auto* switch (HOA). These switches have three possible positions—hand, off, or auto. The symbol for an HOA switch is shown in Figure 3-17.

Because the HOA selector switch is a maintained switch, it operates in the same manner as the local–remote switch example. When it is used in a control circuit, it looks like Figure 3-18.

As shown, the electricity must first flow through the HOA switch before it can energize the coil of the motor contactor. To get through the switch, the switch must be in the hand or auto position. If it is in the hand position, the current can flow

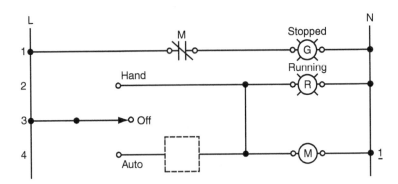

Figure 3-18   HOA motor circuit

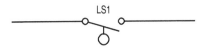

Figure 3-19   Float-operated level switch symbol (closes on rising level)

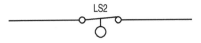

Figure 3-20   Float-operated level switch symbol (opens on rising level)

straight to the motor contactor's coil, causing the motor to run any time the HOA switch is in the hand position. If the HOA switch is in the off position, there is no path for the electricity to get to the contactor's coil, and, therefore, the motor cannot run. If the HOA switch is in the auto position, then there must be some other switch used in the dotted box to allow the current to flow through to the contactor's coil.

What is used depends on what is supposed to start the motor automatically; for example, if a sump pump was installed in a low spot in a plant and was to operate whenever the level in the sump got high, a float-type level switch could be used to start the pump. Then the float could rise up with a high water level and close a switch causing the motor to run a pump and draw down the level. If a switch is used that closes on rising level, it is called a normally open level switch and is shown in Figure 3-19.

The level switch is labeled LS1 to distinguish it from other level switches that might be used. If the switch opens on rising level, it is called a normally closed level switch and is shown in Figure 3-20.

Several other process sensing switches can be used, especially pressure, flow, and temperature. These switches are usually labeled PS, FS, TS, respectively. The symbols for several common process switches are shown in Figure 3-25, located at the end of this chapter, which is a legend sheet showing most of the more common ladder diagram symbols. When switches are used to detect a high value, the letter H is added to the label. For example, LSH indicates a level switch high. Conversely, the letter L is used to label a switch used to detect a low process value. Therefore, an LSL

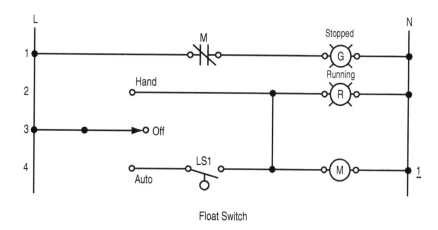

Figure 3-21 Automatic pump control off of a float switch

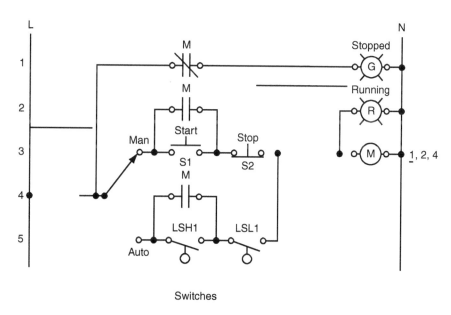

Figure 3-22 Three-wire control using two level switches

would indicate a level switch sensing a low level. If a float-type level switch is used for automatic operation with an HOA switch, the circuit would look like Figure 3-21.

Whenever the HOA switch is in the auto position, the level switch will control the motor. Whenever the level switch is closed, the motor will run. If the HOA switch is in the hand or off position, the level switch will have no effect on the motor.

The example of automatic control used only one level switch and took advantage of the simplicity of two-wire control. Sometimes more than one level switch needs to be used, requiring the use of the three-wire control techniques. For example, the sump pump might be very deep and require the use of one level switch near the top of the sump to start the pump and another level switch near the bottom of the sump to stop the pump (Figure 3-22).

Notice that a selector switch labeled *man–auto* is used here instead of the HOA switch, which was used in the two-wire version. This switch is usually located where the start–stop push buttons are and determines whether the start–stop push buttons or the level switches control the motor. This push button is a maintained selector switch just like a local–remote switch. In the manual position, the current flows through to the left side of the start button on rung number 3 and to the left side of the normally open M contact on rung number 2. When the start button is momentarily depressed, the current will flow through to the M coil and energize it. The energized coil will cause the normally open M contact to close and bypass the current around the start button. This circuit uses an auxiliary contact of the starter contactor (which is just a big relay) to provide the holding contact instead of an extra relay (called CR1 in the three-wire control). The motor will run until the stop button on rung number 3 is momentarily depressed. The current path to the M coil will be interrupted and de-energized. The motor will stop, and the M contact on rung 2 will open up. When the stop button is released, the motor will stay off because the electric latch was reset. The action is identical to that of the three-wire control.

If the man–auto switch is in the auto position, the start–stop push buttons have no effect. Instead, the current must flow through the bottom half of the man–auto switch to the left side of the high-level switch (LSH1) on rung number 5 and to the left side of the normally open M contact on rung number 4. When LSH1 is closed by high water level, the current will flow through it to the low-level switch (LSL1). Because the water is already up to the high-level switch, it is certainly above the lower-level switch (LSL1). LSL1 will already be closed and allow the current to flow through it to the M coil and energize it. The normally open M contact on rung number 4 will close and bypass the current around LSH1. The motor will then run and pump the water down. When the water level recedes below LSH1, it will cause LSH1 to open its contact, but that will not stop the motor because the M contact on rung number 4 is holding the motor on. The motor will continue to run until LSL1 opens on low water level.

In these automatic control examples, the motor cannot be stopped in the automatic mode by any manual button. If an operator at the motor saw a problem with the equipment, there would be no way for the operator to stop the motor without going to the location of the HOA or the man–auto switch. If the HOA or man–auto switch was not at the motor, this would be very unsafe. This unsafe condition can be rectified by putting an extra stop button at the motor. When this is done, the stop button has a flip-over bracket that can be used to keep the button depressed without having to keep pushing on it. The bracket is usually the lockable kind, and the switch is referred to as a lock-out stop (LOS) switch. They are made lockable so that the equipment can be locked in the off state until maintenance staff can be dispatched. This LOS switch should not be confused with the motor disconnect switch (see Motor Disconnects). If an LOS switch is added to the three-wire automatic circuit, it would look like Figure 3-23.

Interlocks

Interlocks are nonmanual control switches activated by an unsafe condition. Interlocks do operate automatically but are not intended to control the motor in normal circumstances. Typically, interlocks stop motors automatically but do not start motors automatically. A few examples of interlocks are

- Motor overloads
- Low level on the suction side of a water pump
- High discharge pressure on a positive displacement pump
- High torque on a conveyor belt

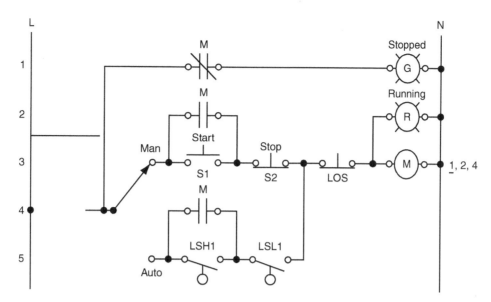

Figure 3-23 Three-wire control using two level switches with lock-out–stop switch

Interlocks are intended to protect human life or equipment, so they are designed to function regardless of whether a manual or automatic control is in use. Therefore, they are wired in a motor control circuit on the line side of other devices; for example, if there is concern that a sump pump will burn out if it is operated when there is no water in the sump. Because this could damage the pump even if it is being operated manually, it would be best to create an interlock for it. To create this interlock, an additional low-level switch is installed in the sump that operates at a level below the low-level switch used to turn the motor off in the automatic mode as described for the three-wire control version. This extra low-level switch would be labeled LSLL; the two Ls on the end to mean low-low (lower than low). Similarly, a high-value interlock might be labeled HH for high-high (higher than high). This LSLL would then be wired into the circuit between the line side of the ladder and the man–auto switch, as shown in Figure 3-24.

The LSLL contact must be closed in order to run the motor in either the manual mode or the automatic mode. Also, the green stopped light will not be illuminated if the LSLL is not closed. Interlocks are often wired in a way that extinguishes all status indicating lights so that if all indicating lights are out, it indicates that an interlock has been tripped. When the green light is wired this way, it is often engraved *ready* instead of *stopped*. The green light is illuminated only if the motor is ready to run. Of course, if the green light was burned out, it would falsely indicate that an interlock was tripped. However, most indicating lights are the push-to-test type, which means that the light's lens is actually a push button that can be pressed to illuminate the light regardless of the current status of the motor. Therefore, simply pushing the light's lens will determine whether the problem is a burned-out bulb or if an interlock has been tripped.

Motor Control Summary

The examples presented have examined two kinds of manual switches, one kind of process switch, and two kinds of relays. There are several other types of switches and other types of relays, such as time-delay relays. These are beyond the scope of this

64 INSTRUMENTATION AND CONTROL

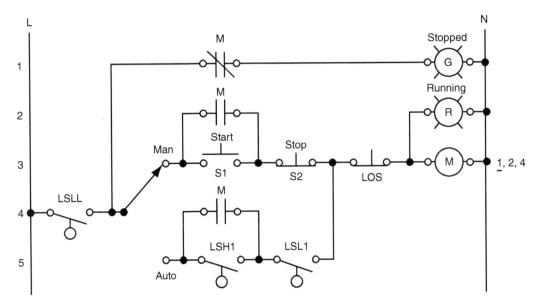

Figure 3-24 Three-wire control using two level switches with lock-out–stop switch and a low-level interlock switch

manual but are shown symbolically on the legend sheet (see Figure 3-25). These can be assembled in thousands of different ways to create the desired control action for any particular situation.

The information presented in this manual is intended to provide the reader with only some of the more common basic logic schemes used to control motors and introduce the reader to ladder diagrams which are used to document the logic schemes. When trying to figure out a motor control diagram, it is important to find a path from the line side of the ladder diagram to the motor contactor's coil to make it run. Any single thing or combination of things that closes a path to the contactor's coil will make it run. And, conversely, anything that interrupts the path will cause the motor to stop.

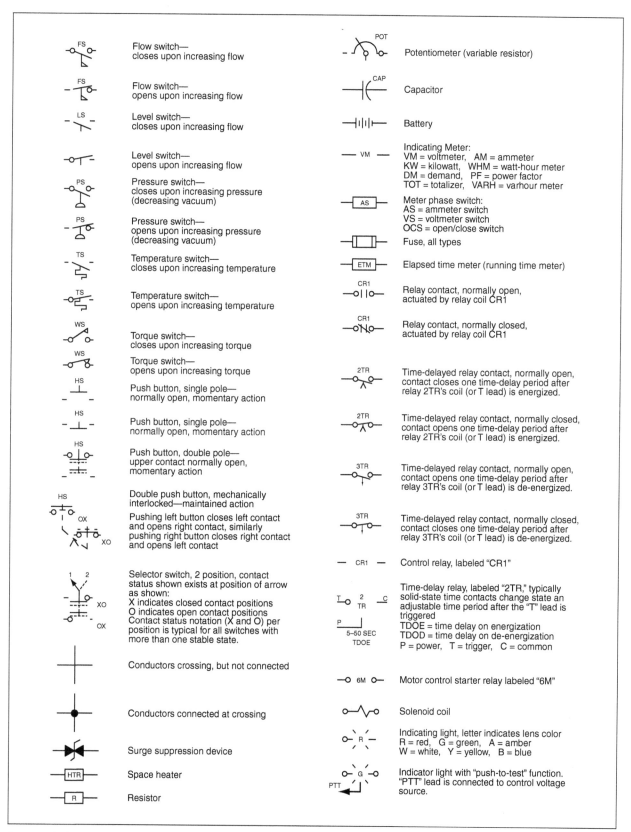

Figure 3-25 Electrical ladder diagram symbol legend

This page intentionally blank.

AWWA MANUAL M2

Chapter 4

Flowmeters

The most significant measurement in water treatment and distribution facilities is flow. Flow values are needed for plant inflow, plant outflow, pacing of chemical feeds, pump output, and splitting streams among tanks and basins. Day-to-day operational decisions and long-term planning are based on the measurements from flowmeters. Operational analysis, performance evaluation, and accounting procedures are derived from plant flows.

Flow can be measured through many types of flowmeters. Recent advancements in flow measurements are based on technological improvements in sensors and transducers, and the application of special purpose microprocessors.

This chapter discusses the most common flowmeters in service in water supply systems. These include the Venturi meter (Venturi), modified Venturis, orifice plate, magnetic, turbine and propeller, sonic, vortex, averaging Pitot, and rotameter. Open channel flowmeters, such as weirs and flumes, are also discussed. Topics covered in this chapter are basic theory, installation, maintenance, advantages, and disadvantages.

No attempt is made to provide complete details for application and selection of flowmeters. The approach, rather, is to provide an introductory overview and guidelines to understanding some of the most common types of flowmeters.

Manufacturers' flowmeter manuals are very comprehensive, and typically include specifications, theory, sizing, handling, installation, power and wiring requirements, operation and maintenance, troubleshooting procedures, parts lists, and costs. The data are accompanied by diagrams, charts, and photographs. Forms to guide the user toward application, sizing, and purchase are always included. Questions about installation and purchase can be answered by contacting the manufacturer directly.

METER CATEGORIES

The meter types described in this chapter are categorized as differential pressure, velocity, fluid-dynamic, or variable area.

The most commonly used flow-measurement devices are the differential pressure flowmeters, also called *head meters*. They are popular because of their flexibility, simplicity, ease of installation, and reliability. The flow velocity of a differential pressure meter is calculated from the difference of two pressures measured in the meter. Differential pressure flowmeters are available in many forms, such as the Venturi, modified Venturi or flow tube, and orifice plate. The measurement of flow is a function of detecting two pressure heads, usually one in the normal pipe size and one in a constricted region, called the throat, within the meter. The flow is proportional to the square root of the differential of the two pressure readings.

In velocity-type flowmeters, the mean velocity of the fluid is calculated or inferred from physical phenomena, such as differential pressure and oscillations caused by bluff body obstructions. Velocity flowmeters include the Pitot, magnetic, turbine, propeller, and sonic flowmeters. In each case, the velocity of the flow is determined. Because the fluid is flowing full in a closed pipe, the flow is calculated from the product of velocity times cross-sectional area. While the Pitot meter is also a velocity-measuring meter, the velocity is determined from the differential pressure of the flow's dynamic pressure and the static pressure.

The fluid-dynamic meter included in this chapter is the vortex-shedding flowmeter. The meter contains a bluff body obstruction in the path of the flow which generates an oscillating motion in the fluid. The oscillations, which are proportional to the flow, are detected and converted to a flow reading output.

In a variable area flowmeter, the volumetric flow is inferred by allowing the area of a flow restriction to vary according to the flow rate, while the average fluid velocity is kept constant. Again, the volumetric flow rate is the product of the velocity times the cross-sectional area through which the fluid passes. Typically, the variable area flowmeter is composed of a tapered tube and a float in a vertical orientation. The upward-flowing fluid velocity pressure balances the gravity and buoyant forces (which are constant) of the float. To create the balance, the float moves to a position in the tapered tube where the velocity in the annulus area generates the equivalent force to offset gravity. The flow is read on a calibrated scale etched into the tube wall.

Other types of differential pressure flowmeters are the open channel flow detectors, such as weirs and flumes. In both cases, the flow is calculated from fluid depth or head which drives the flow.

METER COEFFICIENT OF DISCHARGE

To account for deviations from ideal behavior, a flowmeter performance index is useful. The coefficient of discharge is the most common and is used primarily for differential pressure flowmeters. Manufacturers define a generic meter coefficient of discharge, C, which is directly related to the accuracies found in the manufacturers' published literature.

The coefficient of discharge accounts for deviations from the true average velocity. The ratio of actual flow rate (determined by high-accuracy measurement) to the theoretical flow (calculated from a mathematical model published by the manufacturer) defines the meter's discharge coefficient. A discharge coefficient of unity (1) would indicate a perfect flowmeter, which has never been achieved. However, the closer the coefficient is to 1, the more precisely the meter is constructed and the more accurate the device tends to be under varying operating conditions.

A precise flow calculation frequently depends on much more than the obvious variables entered into the mathematical model. If any flow parameters change, the flow calculation may no longer be valid. Significant changes in fluid temperature, density, and specific gravity may cause errors in flow calculations.

No attempt is made to detail the formulas for the mathematical models of specific flowmeters.

VENTURI FLOWMETERS

- Accuracy: ±0.75 percent of rate
- Repeatability: ±0.25 percent
- Rangeability: 4:1 to 10:1
- Size range: 1–120 in. (25 mm–3 m)
- Head loss: Low
- Relative cost: Medium

The Venturi (Figure 4-1) is the standard of the differential pressure flowmeters. In the Venturi, a defined constriction (throat) within the meter body causes an increase in the velocity of flow at the constriction, resulting in a corresponding decrease in pressure in the throat. The ratio of throat diameter to pipe diameter (where the two pressure measurements are taken) is called the beta ratio. The difference in pressure between the connections upstream and at the constriction (throat) is proportional to the square of the flow. Flow is then calculated from the square root of the measured pressure differential.

To obtain an electrical flow measurement signal, the pressure at the ports is transferred to a differential pressure transducer. The transducer is typically a flexible diaphragm, with a small chamber where the two pressure lines from the meter tube are connected to the two sides. The pressures exert opposing forces on the diaphragm, causing deflection of the sides. To sense the diaphragm's deflection, a strain gauge or a variable capacitance or a variable reluctance device is built into the transducer to generate an electrical signal proportional to the pressure difference. Direct reading of differential pressure is possible using a U-tube manometer or double-bellows gauge.

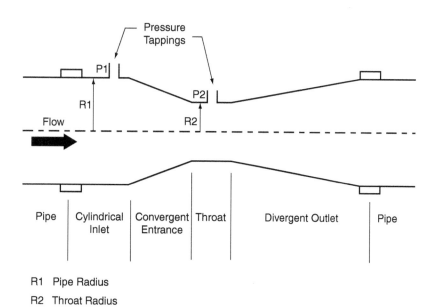

Figure 4-1 The Venturi tube

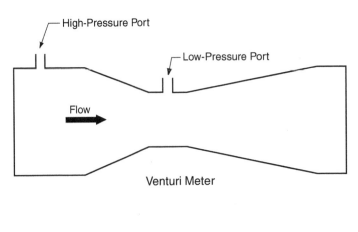

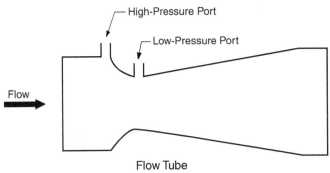

Figure 4-2 Venturi meter and flow tube

The major distinction between the Venturi and modified forms of this device, such as flow tubes and flow nozzles, is that the pressure connections for the Venturi are pure static measurements taken at points where the direction of the fluid flow is not changing (see Figure 4-2). The shape of the approach to the throat conditions the flow pattern in the throat. With hydraulic profile conditioning built into the body inlet structure, shorter lengths of straight pipe may be used on the inlet side when compared to other primary flowmeters. Unrecovered head loss is 10 to 20 percent of the measured differential pressure.

The accuracy of a Venturi is ±0.75 percent of actual flow, independent of additional tolerances due to instrumentation for measurement of the differential pressure. The Venturi has no theoretical range limitations up to the speed of sound. However, the flowmeter, which includes the ΔP device, has definite limits. Some Venturi meters provide a 10:1 range, several 8:1, and many only 4:1, depending upon selection of the ΔP device.

The Venturi meter has a low permanent head loss and does not obstruct the flow of suspended matter. The coefficient of discharge of the classic Venturi is 0.983, which remains constant regardless of pipe or throat size. The Venturi has the most comprehensive documentation regarding its coefficient, with hundreds of thousands of laboratory test points taken in every conceivable piping configuration.

Commonly constructed from cast iron and naval bronze, the Venturi requires the greatest laying length of all differential pressure meters and is, therefore, the heaviest flow tube as well. Fabricated steel ventures and insert-type Venturi meters are also available.

Installation

Factors to consider in the selection process before installation are

- Selecting a meter with a high differential pressure and, therefore, having a small throat size, taking into account the amount of head loss that can be tolerated. A good rule of thumb is to size the tube to provide at least one inch of differential pressure at the minimum anticipated flow. A high differential provides the greatest energy to drive the instrumentation and improve the range and accuracy. Throat size is defined by the beta ratio.

- Reviewing the upstream piping configuration according to manufacturer's requirements. A Venturi is not strongly affected by the downstream configuration except for a slight increase in head loss.

- Considering future expansion of the facility, which may increase required flow.

The installation of a Venturi is critical to the accuracy of the differential pressure measurement. Errors in computing flow may become unacceptable if distorted flow is present. Swirls or vortices in the flow that affect meter accuracy can be produced by a projecting gasket, misalignment, or a burr on a pressure tap.

The Venturi should be installed with its axis horizontal, and the fluid entering the tube with a fully developed velocity profile free from swirls and vortices. In a horizontal installation, the pressure port tappings must not be at the bottom where it is subject to clogging, nor at the top where it is subject to air bubble trapping. The preferred location is on the side in the horizontal plane of the center line. The two pressure lines should be installed with equal length and routed to prevent air or solids accumulation in the connection piping to the differential pressure measuring device.

Maintenance

Because it has no moving parts, the Venturi would normally require less attention than, for example, a turbine meter. However, the differential pressure assembly can have significant piping, fittings, and valves. Lines may clog and corrosion can appear. Periodic disassembly, inspection, and cleaning should be practiced. The pressure sensors, in particular, should be removed and inspected. While the throat of the Venturi must be inspected for debris or deposits, the high fluid velocity usually scours the throat.

Manufacturers provide step-by-step instructions for checking the meter components. The procedures include disassembly, inspection and testing, parts replacement, and reassembly with emphasis on the differential pressure unit and electrical housing. Instructions for zero and span adjustment are also included. A troubleshooting guide may be provided with symptoms, potential sources, and recommended corrective action. To aid the user, illustrated drawings, schematic diagrams, and parts lists are provided in the manufacturer's manual. Figure 4-3 is an example of a troubleshooting guide for a differential pressure transducer.

Troubleshooting—Differential Pressure Transducer

SYMPTOM: High Output
POTENTIAL SOURCE AND CORRECTIVE ACTION

1. **Primary Element**
   - Check for restrictions at primary element.
2. **Impulse Piping**
   - Check for leaks or blockage.
   - Check that blocking valves are fully open.
   - Check for entrapped gas in liquid lines and for liquid in dry lines.
   - Check that density of fluid in impulse lines is unchanged.
   - Check for sediment in transmitter process flanges.
3. **Transmitter Electronics Connections**
   - Make sure bayonet connectors are clean and check the sensor connections.
   - Check that bayonet pin #8 is properly grounded to the case.
4. **Transmitter Electronics Failure**
   - Determine faulty circuit board by trying spare boards.
   - Replace faulty board.
5. **Sensing Element**
   - *See* Sensing Element Checkout Section.
6. **Power Supply**
   - Check output of power supply.

SYMPTOM: Erratic Output
POTENTIAL SOURCE AND CORRECTIVE ACTION

1. **Loop Wiring**
   - Check for adequate voltage to the transmitter.
   - Check for intermittent shorts, open circuits, and multiple grounds.
     NOTE: Do not use over 100 V to check the loop.
2. **Process Fluid Pulsation**
   - Adjust electronic damping pot (4–20 mA DC only).
3. **Impulse Piping**
   - Check for entrapped gas in liquid lines and for liquid in dry lines.
4. **Transmitter Electronics Connections**
   - Check for intermittent shorts or open circuits.
   - Make sure that bayonet connectors are clean and check the sensor connections.
   - Check that bayonet pin #8 is properly grounded to the case.
5. **Transmitter Electronics Failure**
   - Determine faulty board by trying spare boards.
   - Replace faulty circuit board.

(figure continues)

Figure 4-3 Troubleshooting guide for a differential pressure transducer

SYMPTOM: Low Output or No Output
POTENTIAL SOURCE AND CORRECTIVE ACTION
1. **Primary Element**
   - Check installation and condition of element.
   - Note any changes in process fluid properties that may affect output.
2. **Loop Wiring**
   - Check for adequate voltage to transmitter.
   - Check for shorts and multiple grounds.
   - Check polarity of connections.
   - Check loop impedance.
     NOTE: Do not use over 100 V to check the loop.
3. **Impulse Piping**
   - Check that pressure connection is correct.
   - Check for leaks or blockage.
   - Check for entrapped gas in liquid lines.
   - Check for sediment in transmitter process flange.
   - Check that blocking valves are fully open and that bypass valves are tightly closed.
   - Check that density of fluid in impulse piping is unchanged.
4. **Transmitter Electronics Connections**
   - Check to see that calibration adjustments are in control range.
   - Check for shorts in sensor leads.
   - Make sure bayonet connectors are clean and check the sensor connections.
   - Check that bayonet pin #8 is properly grounded to the case.
5. **Test Diode Failure**
   - Replace test diode or jumper terminals.
6. **Transmitter Electronics Failure**
   - Determine faulty circuit board by trying spare boards.
   - Replace faulty board.
7. **Sensing Element**
   - *See* Sensing Element Checkout Section.

Figure 4-3 Troubleshooting guide for a differential pressure transducer—*continued*

Advantages and Disadvantages

The advantages of the Venturi meter are

- Life expectancy of a Venturi with manometer is documented to be greater than 50 years (this may not be true of modern instrumentation in general, although some earlier instruments lasted for 30 years or more); construction materials are well documented for long life.

- Simplicity of construction and no moving parts.

- No sudden change in contour, no sharp corners.

- Relatively high pressure recovery in the outlet cone, yielding a low head loss and substantial power savings for large flows.

- Well documented in the literature as an acceptable type of flowmeter.

The disadvantages of the Venturi meter are

- Larger units are costly to purchase and install.
- Largest and heaviest of the differential pressure meters.
- Differential pressure is not linear with flow rate and requires square root extraction, which reduces rangeability.
- The coefficient of discharge deteriorates for low Reynolds numbers (below 200,000).

MODIFIED VENTURIS

Several flowmeters are derived from the Venturi and operate on the same principle. Classified as differential pressure producers, they include flow tubes or low-loss (proprietary) flow tubes. The object of any modification to the Venturi is to achieve shorter laying length, less cost, and higher head recovery. The length of the throat in the flow tubes is much shorter than in Venturi meters, which may result in a less stable flow pattern through the throat and reduce accuracy.

Flow Tubes

Although similar to the Venturi, the flow tube has a relatively short transition section from the inlet to the throat (see Figure 4-2) and a recovery cone gradually increasing in diameter. The flow tube causes significantly less pressure drop than the orifice plate flowmeter, but it is costlier. With its shortened inlet section, the flow tube is usually less expensive than the Venturi for the same line size. The body structure does not provide hydraulic profile conditioning and, therefore, is more sensitive to upstream flow disturbances.

A very low head loss, lower than that for a comparable Venturi with the same beta ratio, has been achieved by the low-loss flow tube. The head loss of the tube is a percentage of the differential pressure, based on throat size.

Flow tubes are typically proprietary designs, and the user should use caution and apply the meter within the calibrated ranges. Supporting application data are generally available only from the manufacturer.

Because the materials of construction for the flow tube are the same as those for the Venturi, long life may be anticipated. The accuracy is ±1 percent of the flow rate, and the range, like the Venturi, is limited only by the associated instruments. The coefficient of discharge also deteriorates significantly at low Reynolds numbers.

Insert Flow Tubes

Another variation of the Venturi meter is the insert flow tube. This unit is convenient because of its short laying length, single mounting flange with no inlet cone upstream of the flange, and low head loss. Both the high- and the low-pressure metering taps are built into the flange.

ORIFICE PLATE FLOWMETERS

- Accuracy: ±1 to ±2 percent of full scale
- Repeatability: ±0.25 percent
- Rangeability: 4:1
- Size range: All

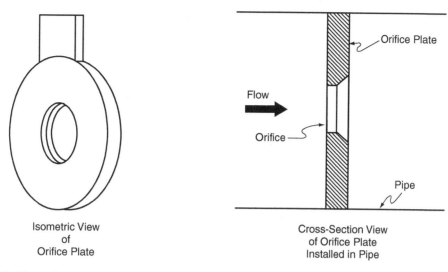

Figure 4-4 Orifice plate

- Head loss: Medium

- Relative cost: Low

The orifice plate flowmeter consists of a thin plate with a hole in it (Figure 4-4). The standard plate is a circular disc, fitting snugly in the pipe, usually of stainless steel, 1/8–1/2 in. (3–13 mm) thick, and containing an orifice with a sharp leading edge. The hole is usually concentric with the pipe into which the plate is inserted perpendicular to the axis of flow. The plate is typically installed in a pipeline between two flanges, and the differential pressure across the plate is measured. Because of its simplicity, low cost, ease of installation, and reasonable accuracy, the orifice plate is among the most common primary elements for measuring flow.

Most orifice plate meters for clean water are made with a circular orifice concentric with the pipe. In special applications, notably for flowing solids, the orifice may also be eccentric or segmental. Passage of entrained solids is permitted if the hole is tangent to the inside surface of the bottom of a horizontally laid pipe.

As in a typical differential pressure flowmeter, the pressure upstream (approximately one pipe diameter from the plate) is compared with the pressure downstream where the flow converges to the point of narrowest stream flow. The narrowest section of flow is called the *vena contracta*, and is taken to be one-half pipe diameter downstream from the plate. These points define where to locate the pressure taps.

Numerous eddies form downstream from the plate between the pipe wall and the *vena contracta*, causing kinetic energy to be dissipated as heat. This dissipation accounts for the medium to high head loss associated with the orifice plate meter.

The orifice plate size can be fabricated to accommodate any pipe size. Inherent accuracy, independent of pipe diameter errors and the differential pressure sensor, may be ±1 to ±2 percent of full scale.

Orifice plates are also available as part of integrated pipe assemblies, which include the plate, pipe length, pressure taps, and, in some cases, the differential pressure sensor and signal transmitter.

Installation

The orifice plate is usually mounted between a pair of flanges. Manufacturers may extend the plate to include a tab above the edge of the pipe flange. The tab, suitable for a nameplate, may contain pertinent data on the specific installation and may identify the upstream side. To prevent errors in flow measurement, the gaskets should not protrude across the plate face beyond the inside pipe wall. Typically, the orifice plate also requires a straight run of smooth flow before and after the plate. Pressure taps must be installed perpendicular to the pipe wall. For horizontal pipe runs, the pressure taps should be in the horizontal plane of the pipe center line. Burrs and intrusions at the taps must be removed.

Maintenance

As a differential pressure flowmeter with no moving parts, the orifice plate meter requires maintenance similar to that for the Venturi. The orifice should be visually checked periodically to be sure its dimensions and sharp leading edge are unaffected by the flow. Degeneration of the sharp edge can result in significant errors in the measurement. Solids may collect behind the plate at the bottom and gases may be trapped at the top. These may be easily cleared by removal of the plate. Some manufacturers provide special mounting devices to allow the orifice plate to be inserted and removed without interrupting the flow, which simplifies maintenance.

The pressure taps should also be examined for possible obstruction. The differential pressure measurement assembly should be periodically checked.

Advantages and Disadvantages

The advantages of orifice plate flowmeters are

- Lowest-cost differential pressure meter
- Economically manufactured to very close tolerances
- Easily installed and replaced
- No moving parts

The disadvantages of orifice plate flowmeters are

- High permanent pressure loss through orifice plate can cause significant power costs
- Volume flow is nonlinear
- Rangeability is lower than with linear-output flowmeters
- Long upstream and downstream straight pipe runs are required

MAGNETIC FLOWMETERS

- Accuracy: ±0.5 percent of rate
- Repeatability: ±0.25 percent
- Rangeability: 10:1 to 30:1
- Size range: 0.1–120 in. (2.5 mm–3 m)
- Head loss: None
- Relative cost: High

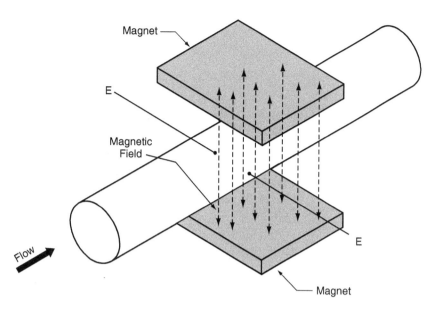

Figure 4-5　Magnetic flowmeter

In a magnetic flowmeter (Figure 4-5), a magnetic field is generated around an insulated section of pipe. Water passing through the magnetic field induces a small electric current that is proportional to the velocity of the water flow. The electric current is measured and converted to a numerical indication of flow. The voltage generated is approximately 0.05 V/ftm^3/sec for the velocity of the fluid, and the output is essentially linear. The liquid moving through the meter must have sufficient ion content to provide a minimum conductivity. Water treatment plant flows meet this conductivity requirement.

The meter's accuracy is ±0.5 percent of the actual flow rate; it can be used over a velocity range of 0.1–35 ft/sec (0.6–11 m/sec). Although the magnetic flowmeter can sense very low velocity flows, the accuracy is reduced at low velocities. The range of a magnetic flowmeter is 10:1 to 30:1. Typical allowable velocity range in water treatment is from 0.1 ft/sec (0.03 m/sec) minimum to 10 ft/sec (3 m/sec) maximum.

The head loss across a magnetic flowmeter is the loss that is caused by the length of pipe forming the body of the meter. Head loss would be increased if the pipe diameter had to be reduced and a smaller size meter incorporated in order to raise the flow velocity in the meter to an acceptable level. Sizes of magnetic flowmeters range from 0.1 to 120 in. (2.5 mm to 3 m) in diameter.

A magnetic meter has no constriction in its cross-sectional area, so the flow profile is not affected. Therefore, whatever profile error may exist in the piping is carried through the metering area. However, area averaging makes the magnetic flowmeter less sensitive than other types of flowmeters to profile changes. The magnetic meter does not have a discharge coefficient definable by its hydraulic shape. Consequently, all magnetic meters should be wet-calibrated by the manufacturer.

Some of the advantages for this type of meter include its shorter laying length and lighter weight than a Venturi meter. The solid-state electronics that generate the flow signal have a long life.

78   INSTRUMENTATION AND CONTROL

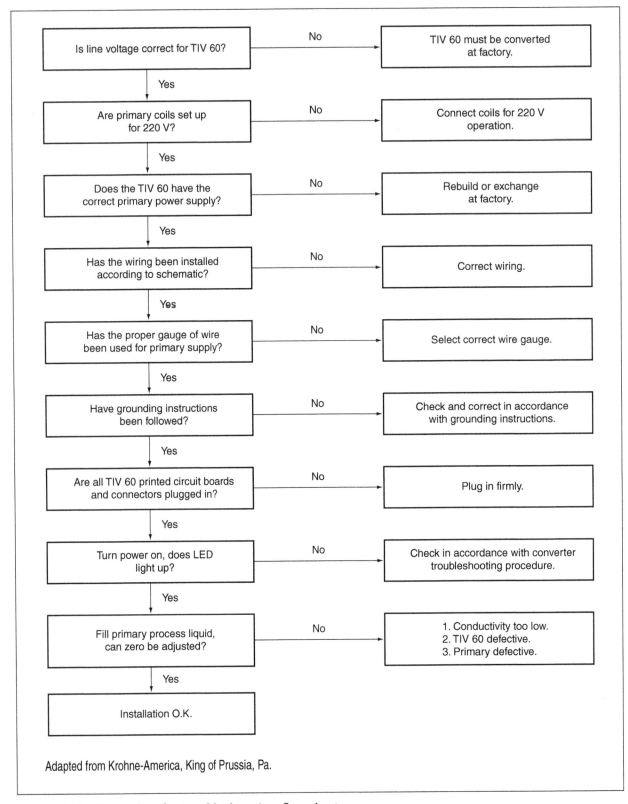

Figure 4-6 Example of a troubleshooting flowchart

Installation

A magnetic flowmeter should be installed with the electrodes located at the ends of a horizontal diameter rather than a vertical diameter. This will ensure continued electrode immersion even when air bubbles are present in the water flow. The meter must always run with full flow.

The following installation steps should be taken in order for the meter to function properly:

- Consider line size against minimum velocity.

- Provide suitable upstream piping length, as recommended by the meter manufacturer.

- Ensure proper grounding isolation between the meter body and the pipeline to avoid transient voltage interference.

- Provide 120 VAC power input at about 100 W (models are available for 220 VAC and 24 VDC).

- Use electrical bonding strips to bypass the cathodic protection currents around the meter if the meter is installed in a pipe that is part of an electrogalvanic-corrosion prevention system.

- Internal lining is critical to the application. Liners range from low-cost polyurethane to high-cost Teflon. The manufacturer should be consulted for the liners most suitable to the application.

Maintenance

Troubleshooting and maintenance procedures recommended by manufacturers for magnetic flowmeters are frequently documented on logic flowcharts. These charts include functional statement blocks and *Yes–No* arrows that indicate the next step in the checking sequence. By using the chart, the meter repair person is directed to the problem sources, diagnostics, and recommendations for remedial action. Flowcharts may be provided for meter installation and primary, converter, and power supply checks. Figure 4-6 is an example of a troubleshooting flowchart used to check meter installation.

Some manufacturers provide special equipment for testing, troubleshooting, and recalibrating meters. Instructions and detailed diagrams are provided for disassembly and replacement of components. If a meter is considered unrepairable in the field and the instructions indicate a warning against tampering with the sealed portion of the unit, no routine maintenance is recommended. Service or installation problems should be referred to the manufacturer's service or field office.

Electrode Cleaning

Deposits and incrustations, including calcium carbonate, on the meter's electrodes can impair accuracy, unless the deposits have the same conductivity as the fluid. These deposits can cause variable or high resistance between the electrodes, thereby introducing errors. Methods for cleaning the electrodes vary among the manufacturers. The following are examples of what may apply to magnetic flowmeter electrodes:

- Some electrodes may be removed for inspection, cleaning, and replacement without removing the meter from the line.

- Conical electrodes are available that extend into the flow stream where the scouring effect of the liquid velocity is more likely to inhibit coating.

- Continuous ultrasonic vibrations will prevent deposition of particulate and crystalline materials on the electrodes; however, this is not effective for grease deposits.
- Bypass piping to maintain flow may be included in the installation to allow for periodic inspection and cleaning of the meter's inner wall.

Recent advances in electronic signal conditioning have significantly reduced the effects of electrode fouling. Manufacturers now state that among the newer meters cleaning is not necessary, except under unusual circumstances.

Advantages and Disadvantages

The advantages of magnetic flowmetering devices are

- No obstruction to flow
- Minimum effective head loss, essentially that of the straight pipe equivalent of the meter (unless meter spool size is reduced from pipe size and thereby causes head loss); magnetic flowmeters are highly suitable for applications where low head loss is essential.
- Available over a wide range of sizes from 0.1 to 120 in. (2.5 mm to 3 m) in diameter
- Bidirectional, therefore, suitable for measuring reverse flows
- Output signal is linear with flow velocity.
- Variations in fluid density, viscosity, pressure, and temperature have little effect on performance.
- Suitable for short runs of straight pipe because unless very severe, upstream nonsymmetrical flow patterns and flow disturbances do not seriously affect the flow measurement.
- Capable of measuring very low flows.

The disadvantages of magnetic flowmetering devices are

- Metered liquid must have an electrical conductivity of 20 µS/cm or greater (this is not a problem with finished drinking water).
- For smaller pipe sizes, the meters become relatively bulky and expensive.
- High accuracy is expensive, and each meter must be individually calibrated in a water test circuit.
- The meter is sensitive to the geometry and electric properties of the flow tube and magnetic core, and to variations in the coil supply current.

TURBINE AND PROPELLER FLOWMETERS

- Accuracy: ±0.5 to 2 percent of rate
- Repeatability: ±0.02 percent
- Rangeability: 10:1 to 50:1
- Size range: $3/16$–24 in. (5–600 mm)

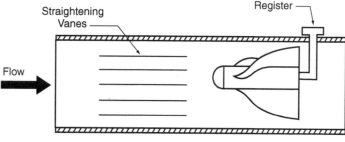

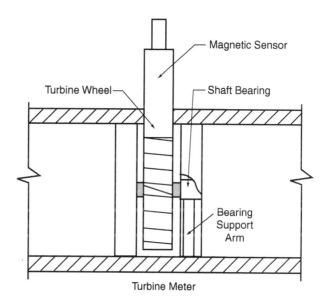

Figure 4-7 Propeller and turbine meters

- Head loss: Moderate
- Relative cost: Medium

In turbine and propeller flowmeters, flowing water strikes rotor blades, which rotate at a rate proportional to the flow velocity. The rotor (turbine wheel) of a turbine meter generally fills the cross section of the pipe and is mounted to spin freely between two central bearings supported in the pipe wall (Figure 4-7). The rotor (propeller) of a propeller meter is mounted on bearings at the downstream end of the pipe and does not fill the meter cross section. The tapered propeller nose faces upstream into the flow and is mounted to spin freely in line with the pipe axis.

Within given limits of flow rate and fluid viscosity, the rotor speed and volumetric flow rate maintain a linear relationship. A magnetic proximity sensor in the meter transduces the rotor velocity to an equivalent frequency signal. Therefore, the rotation of the rotor blades causes a known number of cycles per unit volume. The meter coefficient is the calibration K factor, which is a known number of pulses for a given volume measured. The K factor is typically constant over a 10:1 flow range within a linearity tolerance of ±0.25. For identical meters, the K factor may vary as a

Troubleshooting—Turbine Meter

INTRODUCTION

Turbine meter system malfunctions are usually restricted to two areas: electrical/electronic or mechanical.

When a malfunction or an apparent malfunction occurs, the electrical and electronic systems should first be thoroughly checked in accordance with the manufacturer's recommended procedures prior to checkout of the turbine meter. Only when the source of the malfunction cannot be found in the electrical or electronic systems should the turbine meter be inspected.

SYMPTOM: Fluid Delivery Greater Than Indicated on Totalizer

NOTE: Verify that the proper "K" factor value is entered in the electronic readout device.

1. **Possible Cause**—Foreign material collected on rotor or bearings. When foreign material collects on or in the bearings or if material wraps around the rotor (such as strands of PTFE tape), the angular velocity of the rotor will be reduced. This allows more fluid to pass through the meter than the pulse train indicates.

   Corrective Action—Remove the meter from the line and visually inspect internally. If foreign material is present, remove the material. If no foreign material is found, disassemble the meter in accordance with the instructions of Paragraph 4-3. Clean parts according to Paragraph 4-1. Reassemble and reinstall the meter in the line. If no foreign material is found and cleaning does not eliminate the problem, check bearing wear as performed in the next paragraph.

2. **Possible Cause**—Excessive bearing wear. Excessive bearing wear will lower the angular velocity of the rotor, and rotation may stop completely. Effect is the same as with foreign material.

   Corrective Action—Replace bearings or bushing assembly and journal. Recalibrate as required in accordance with Paragraph 4-2.

SYMPTOM: Fluid Delivery Less Than Indicated on Totalizer

NOTE: Verify that the proper "K" factor value is entered in the electronic readout device.

1. **Possible Cause**—Ground loop in electrical circuit.

   Corrective Action—See manufacturer's installation manual.

2. **Possible Cause**—Gasification of liquid in meter. ITT Barton recommends that a back pressure be applied to the system. This back pressure should be twice the magnitude of the net pressure loss through the meter plus twice the magnitude of the vapor pressure of the metered liquid. If this back pressure is not maintained, gasification within the meter can occur, causing an over-spin of the rotor. This will indicate a greater-than-actual delivery and cause damage to the bearings.

   Corrective Action—Provide a back pressure at the meter by using accepted design means and practices.

3. **Possible Cause**—Entrained gas or bubbles in metered liquid. If the fluid has a significant amount of entrained gas or gas bubbles that are released in the reservoir, over-registration can occur. Turbine meters measure the actual volume of the liquid. If a unit volume is expanded by entrained gas or bubbles, this expansion will be registered. Gas bubbles may cavitate the rotor and produce effects similar to those caused by gasification of liquid in the meter (see previous paragraph).

   Corrective Action—Eliminate entrained gases or bubbles by use of an air eliminator or other acceptable method.

Adapted from *Turbine Meter Troubleshooting Procedures* (ITT Barton, City of Industry, Calif.)

Figure 4-8 Troubleshooting procedures for turbine meter

result of manufacturing tolerances. Therefore, each meter should be calibrated for its own specific K factor value.

According to meter manufacturers, turbine meter accuracy varies from ±0.5 percent to ±2 percent of flow. Manufacturers of propeller meters agree on an accuracy of ±2 percent of actual flow. At low-flow rates, accuracy is reduced to ±5 percent. The rangeability in larger meter sizes varies with a good repeatability at 10:1 and can go up to 25:1.

The orientation and configuration of the meter blade profile are important to the application of the turbine meter. Straight-bladed meters may be less affected by variations in velocity, while helical-bladed meters are generally less affected by variations in viscosity.

Installation

Turbine and propeller flowmeters may be installed with a strainer to prevent solids from interfering with the rotor mechanism. Because these meters are affected by upstream configurations that cause swirls or velocity fluctuations, manufacturers frequently provide or recommend built-in straightening vanes upstream in the pipe. These are installed to minimize the effect of profile irregularities and to smooth flow entering the meter. Straight pipe lengths of at least five pipe diameters upstream and two pipe diameters downstream are recommended.

When selecting a meter for a specific application, care should be taken to ensure that the maximum flow rating will not be exceeded, except for short periods. Running over the maximum speed for extended periods will increase bearing wear and shift the meter's K coefficient.

Location near a point of chemical injection should be avoided, and the electronic mechanisms that generate the pulses should be protected from electromagnetic influence.

Maintenance

A turbine or propeller flowmeter will yield long life, provided periodic mechanical maintenance is performed. Factory maintenance programs are available for meter testing, maintenance, and recalibration. Periodic inspection, calibration, and service should be performed at least once a year.

Propeller meters are generally manufactured for ease of disassembly and extraction of the metering unit from the pipe body. According to many manufacturers, no special tools are required for maintenance. When the metering unit is removed, a cover plate is commonly available for installation to continue line service. During service and disassembly, the complete metering assembly should be examined for wear and corrosion. Parts should be cleaned and worn or damaged parts replaced. Troubleshooting procedures are provided by manufacturers, covering problems, causes, and recommended corrective actions. An excerpt from the troubleshooting procedures for a turbine meter is shown in Figure 4-8.

Advantages and Disadvantages

The advantages of turbine and propeller flowmeters are

- The very large sizes have good repeatability and can be very accurate when calibrated periodically.

- Output flow signal is directly proportional to a pulse train with high linearity over a broad range of flow rates at least 10:1 for large meters.
- Head loss is low to moderate, decreasing with larger sizes.
- Meter size is the same as the diameter of pipe in which it is installed.
- Flow is not blocked if the meter seizes up.

The disadvantages of turbine and propeller flowmeters are

- Systematic mechanical maintenance and lubrication are required.
- Wear or fouling of meter surfaces will gradually deteriorate the calibration, making periodic recalibration necessary to maintain high accuracy. Corrosive liquids, liquids of poor lubricating quality, and liquids with a high proportion of suspended solids will cause bearing problems.
- Calibration factor is sensitive to changes in viscosity of the flowing liquid.
- Sensitivity to flow disturbances and swirls.
- Bearing friction is detrimental to performance of smaller meters.
- Accuracy decreases significantly at low flows.

SONIC FLOWMETERS

- Accuracy: ±1 to ±2.5 percent of rate
- Repeatability: ±0.25 percent
- Rangeability: 20:1
- Size range: 0.125–120 in. (3.1 mm–3 m)
- Head loss: None
- Relative cost: Low

In a sonic (or ultrasonic) flowmeter, a pair of transceivers (transmitter–receiver) are positioned diagonally across the meter body, as shown in Figure 4-9. The transceivers transmit and receive an ultrasonic pulse in the direction of flow, followed by a return pulse against the direction of flow. In a flowing liquid, the speed of the pulse directed downstream is increased by the speed of the stream; when directed upstream, the speed of the pulse is slowed by the stream flow. The time difference

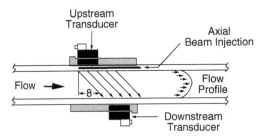

Figure 4-9 Ultrasonic time-of-flight flowmeter

between the two pulse transmissions through the stream is a function of fluid velocity and, by computation, the rate of flow. The transit-time difference and a given value of pipe diameter are converted by a microprocessor circuit in the meter to a standard output signal for volume flow. This sonic meter, used to measure the flow of clean water, is known as a *time-of-flight*, *transit-time*, or *through-transmission* meter. The sonic meter accuracy is ±1 percent of rate over a 10:1 range. Accuracy must be checked in the manufacturer's literature, because some manufacturers cite accuracy in terms of percent full scale.

Because sonic flowmeters use microprocessor circuits to generate and record flow, special digital features can be incorporated into the unit. Light-emitting diodes (LEDs) can indicate circuit operation or alarms and provide digital (numerical) readouts. The user can select output units, ranges, totalizing, and the type of signal transmission.

The head loss of the meter is no more than that of an equivalent length of pipe. In cases of low-flow rates, a less-than-line-size meter may be used, which will create some additional head loss.

Because there is no constriction, as in a Venturi or flow tube, there may be no flow conditioning. Therefore, any profile error caused by the piping configuration is carried through the meter and picked up by its velocity-sensing system. However, flow straighteners may alleviate the problem of turbulence. Some sonic meters use a single-path sensor system that requires more careful layout of upstream piping. To minimize profile sensitivity, some meters have multiple-path sensing to average out the fluid profile error and, therefore, increase metering accuracy.

Another type of ultrasonic flowmeter is the Doppler, which requires only one transceiver to send pulses into the flow. This meter depends on particles of solids or entrained bubbles in the fluid to reflect energy pulses back to the transceiver. Consequently, this type of meter is not suitable for clean, potable water.

Installation

A sonic meter requires a minimum of ten pipe diameters of straight run upstream and five pipe diameters downstream for proper performance. The length of a straight run may differ with varying piping configurations and manufacturers. The meter should not be located near a point where there is a sudden pressure drop that might release minute quantities of gas from the liquid. The meter cannot function accurately when gas bubbles are present. Piping should be isolated from noise and vibration that might interfere with the sound propagation of the meter. The meter requires a power source and may use from 8 to 50 watts of power.

Sonic meters are available in two forms: a spool piece with integral transceivers or a transceiver assembly for clamp-on mounting outside an existing pipe.

Maintenance

Maintenance of sonic flowmeters is minimal. They should have a long operational life because of obstructionless flow, solid-state circuitry, and no moving parts. Built-in electronic self-diagnostics generate circuit tests and display functions, and plug-in modular construction permits rapid replacement of defective parts.

Maintenance procedures are executed through the microprocessor, checking and adjusting the 4 and 20 mA output levels, and the level of the transmitted and received signal. A low signal level may indicate incorrect transducer installation, obscured sonic path (bubbles, solids), sedimentation, or cabling fault.

Advantages and Disadvantages

The advantages of sonic flowmeters are

- No obstruction to flow; therefore, no head loss
- Not restricted to use with conductive liquids (as are magnetic flowmeters)
- Clamp-mounted meters do not jeopardize pipe wall structure.
- Clamp-mounted meters do not interrupt process flow during maintenance or replacement.
- No mechanical moving parts
- Linear output over a wide range
- Adaptable to a wide range of pipe diameters
- Accurate readings of flows down to 0.1 ft/sec
- Low installation and operating costs
- Bidirectional flow is allowable.

The disadvantages of sonic flowmeters are

- Sensitive to change in fluid composition
- High solids content or entrained bubbles distort and block propagation of sound waves.
- Measures mean velocity across a diameter, which is not the same as the weighted mean velocity.
- Sensitive to flow-velocity profile; accuracy can be impaired by changes in pipe wall roughness and by changes from laminar to turbulent flow.
- Accuracy impaired by upstream and downstream flow disturbances, such as elbows and valves, which affect the velocity profile
- Positioning of the opposing transceivers is critical to ensure signal interception.
- In clamp-mounted use, the presence of sound absorptive or scattering scale or coating on the inner walls of the pipe may prevent the meter from working (this is not true when transceivers are mounted through the wall on a spool piece).
- Sensitive to noise and vibration

VORTEX FLOWMETERS

- Accuracy: ±0.75 percent of rate
- Repeatability: ±0.15 percent
- Rangeability: 40:1
- Size range: 1–10 in. (25–250 mm)
- Head loss: Medium
- Relative cost: High

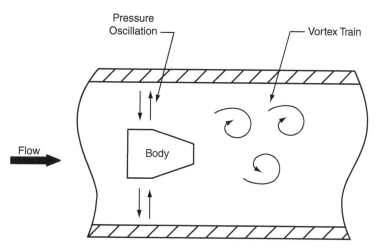

Figure 4-10 Vortex flowmeter

In a vortex flowmeter, a nonstreamlined or bluff body, the vortex-shedding element, obstructs and splits the flow through the pipe forcing two streams around the barrier and creates (or sheds) vortices downstream in the flow. These vortices are caused by the swirling of the fluid into the low-pressure area behind the body (see Figure 4-10). The shedding vortices alternately rotate in opposite directions with the spacing between them proportional to the fluid velocity. This also creates an oscillating pressure variation from side to side of the immersed vortex-shedding element.

Numerous methods are available for measuring the frequency of the vortex train or the frequency of the pressure oscillations. In all cases, external electronics convert the frequency signal into a standard analog value proportional to the flow velocity or into a pulse train suitable for input to a totalizer.

The accuracy of vortex flowmeters is ±0.75 percent of rate with a repeatability of better than ±0.15 percent. Maximum flow rate is approximately 15 ft/sec (5 m/sec) with a range of 40:1. The flow range is a function of pipe size, being dependent on the Reynolds number and on cavitation in the pipe.

Head loss is somewhat higher than that of an unobstructed pipe because of the presence of the vortex element. The added loss is equivalent to about 4 psi (28 kPa) at maximum flow.

The frequency/flow characteristic in the pipe is a function only of the shape of the body. Consequently, a generic coefficient can be used for all meters having the same body profile, regardless of pipe size.

Pipe sizes used with vortex flowmeters range from 1 to 10 in. (25 to 250 mm), and larger sizes can be used. The maximum size of pipe depends on the pulse frequency per unit volume. This limit exists because, in larger pipes, the full-scale frequency may be too low to enable the signal-conditioning electronics to make acceptably accurate measurements.

The vortex meter has no moving parts and has only the spool piece and body exposed to the fluid. It is typically constructed of stainless steel, resulting in relatively low maintenance requirements.

Installation

The vortex meter requires a fully developed flow profile, which means that upstream disturbances must be minimized. An upstream straight pipe length of at least

88 INSTRUMENTATION AND CONTROL

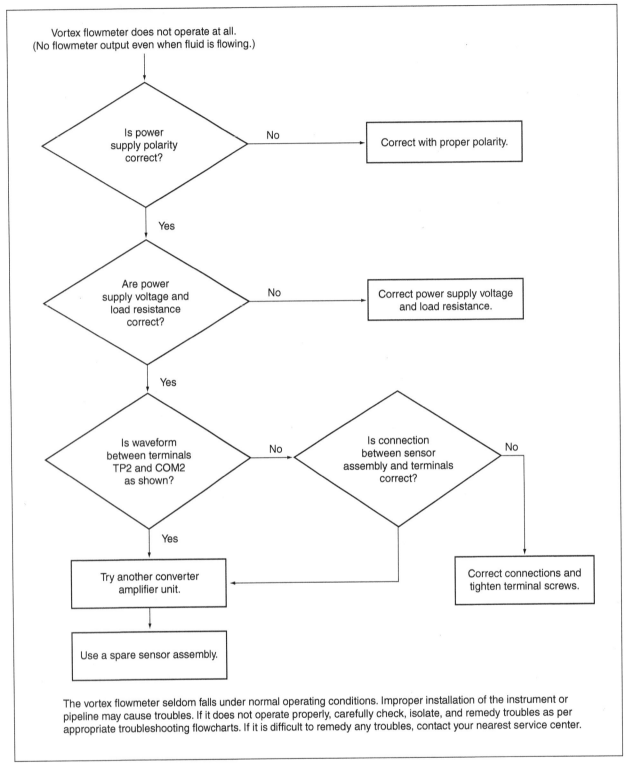

Figure 4-11 Vortex flowmeter troubleshooting guide

ten diameters is desirable, and flow straighteners may be needed where severe disturbances are present. A length of five diameters of straight pipe downstream should be used to minimize the effects of disturbances on the vortex train. Severe vibration of the pipe caused by noisy pumps and valves and the continued presence of bubble streams can affect meter accuracy by introducing false signals into the sensing elements. These problems must be considered before installation of the meter.

Maintenance

If a failure occurs, flowcharts of troubleshooting instructions and related diagrams are provided by manufacturers to isolate and remedy the problem. An example of a flowchart for troubleshooting is shown in Figure 4-11.

To disassemble a meter, the parts are usually removable and replaceable in the field, according to detailed procedures found in the instruction manuals. These may include the vortex-shedding element, the sensor assembly, and output signal electronics. Often the electronics and sensor can be replaced without interrupting pipe flow.

Because it has no moving parts, the installed fixed assembly of meter spool and vortex shedder requires virtually no maintenance. This assumes that installation is executed according to manufacturer instructions regarding orientation, alignment, piping connections, ambient conditions, power connections, wiring, and flow range applications.

Detailed instructions for maintenance usually refer to electronic adjustments, such as zero, span, noise balance, and minimum measurable velocity. However, these adjustments, while explained in the meter manual, are set in the factory and should not be altered after proper installation.

Advantages and Disadvantages

The advantages of vortex flowmeters are

- Low head loss
- No moving parts
- Long-life construction
- High turndown ratio or rangeability (the ratio of the maximum design flow rate to the minimum design flow rate)
- Simplicity in design and installation

The disadvantages of vortex flowmeters are

- Sensitive to flow-profile distortion
- Affected by pipe vibration and bubble streams
- Not bidirectional
- Limited range of pipe size
- Cannot measure flow below low-flow cutoff velocity

AVERAGING PITOT FLOWMETERS

- Accuracy: $\pm \frac{1}{2}$ to ± 5 percent of full scale
- Repeatability: 0.5 percent
- Rangeability: 4:1

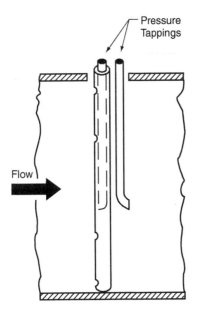
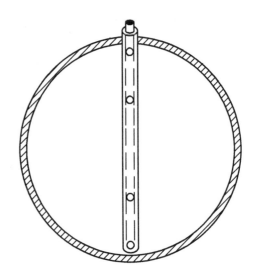

Figure 4-12 Averaging Pitot flowmeter insertion tube

- Size range: ½–96 in. (13 mm–2.4 m)
- Head loss: Low
- Relative cost: Low

The averaging Pitot flowmeter consists of an insertion tube, or probe, that is placed along a diameter through the pipe cross section (Figure 4-12). Multiple ports face upstream into the flow to provide sampled pressures at selected points along the vertical pipe diameter. The multiple pressures are sensed as an interpolated average by the internal tube to provide an averaged pressure over the pipe cross section. A port or ports facing downstream register static pressure. The device produces a differential pressure between the averaged velocity head ports and the static head port(s). As with the Venturi flowmeter, the fluid velocity is proportional to the square root of the differential between the resultant upstream pressure and the static pressure.

Accuracy for a Pitot meter is ±½ percent to ±5 percent of full scale. The meter's range is greatly limited by the sensitivity of the differential pressure sensor and the results obtainable for low differential. The meter, by nature, is velocity-profile sensitive. Placing it in different planes at the same point in a pipe may result in different readings. Laboratory calibration is not valid for field conditions unless the pipe configuration is duplicated.

According to manufacturers, the shape of the insertion tube generally will cause most foreign material to flow around the probe rather than accumulate on it. Ordinarily, material that does affect the probe does not significantly affect meter performance unless, in extreme cases, the ports are completely obstructed or buildup changes the outside shape.

Installation

The insertion tube must be installed into the pipe according to the manufacturer's instructions. Deviation from perpendicular to the axis of the pipe in any direction will affect the sensed pressures. Misalignment of the tube from the diameter beyond 3° in the pipe cross-sectional plane or 5° upstream or downstream out of the cross-sectional plane will cause significant error. Rotation of the tube beyond 3° from a strict upstream–downstream orientation will also cause errors in the sensed pressure.

The correct size insertion tube must be installed in the correct size pipe. The location of the ports on the insertion tube is based on the pipe's inside diameter and wall thickness. Installed in the proper size line with the proper fittings, the tube's sensing ports will be at their correct locations, and the meter will respond with the designed accuracy. If the tube is inserted in an incorrect line size, it can be expected to provide a repeatable signal, but must be recalibrated to yield accurate flow measurements.

The location of tube ports is based on a fully developed turbulent flow with a velocity profile that is consistent across the pipe in all directions. The averaging of pressures will not be correct in an inconsistent flow profile, and errors in flow measurement will result. Upstream devices, such as valves, elbows, and pipe diameter changes, influence the flow profile. Therefore, as with other meters, sufficient lengths of pipe must be provided upstream of the insertion tube to allow the turbulent flow profile to develop. Flow straighteners may be used to reduce the necessary length of straight pipe upstream. Tables of recommended upstream and downstream straight pipe lengths are provided by manufacturers. Upstream lengths vary from 7 to 24 pipe diameters without straighteners and 3 to 9 pipe diameters with straighteners. Downstream straight pipe lengths vary from 3 to 4 pipe diameters.

Maintenance

The probe should be removed and cleaned periodically. Probe removal does not require shutting down the system. Sensing ports and internal passages can be cleaned using external pressure without being removed. Manufacturer's recommendations should be followed in cleaning the probe. The sensor design will handle most flow conditions without clogging. Nevertheless, if the fluid is contaminated, periodic purging of the internal passages may be necessary. Designs can be provided to facilitate purging.

Advantages and Disadvantages

The advantages of averaging Pitot flowmeters are

- Removable without shutting down system (hot tap option)
- Nonconstricting design produces low head losses
- Meter cost is low
- Easily installed at any time by making a wet tap in the pipeline
- Materials of construction and the nature of the parts provide for long life

The disadvantages of averaging Pitot flowmeters are

- Any leaks in the instrument lines or connections will significantly affect the meter accuracy, because the flow measurement depends on the comparison of two pressures generated by the flow past the device, and the pressure change may be small
- Potential for calibration error exists due to misalignment of the insertion tube
- Requires square root extraction

VARIABLE AREA FLOWMETERS

- Accuracy: ±0.5 percent of rate to ±10 percent full scale
- Repeatability: ±$\frac{1}{4}$ percent
- Rangeability: 10:1
- Size range: $\frac{1}{16}$ to 12 in. (2.0 to 300 mm)
- Head loss: Low
- Relative cost: Low

The variable area meter, also known as the rotameter, is a variation on the differential pressure meter. The area through which the liquid flows is permitted to vary so that a constant differential pressure is maintained. This is achieved through the construction and application of the meter. The basic elements of the rotameter are a tapered (conical) tube, vertically oriented, and a cylindrical float, free to rise and fall in the tube. The metering tube is mounted in a vertical position with the tapered end at the bottom where the fluid to be measured enters and rises, filling the tube, passing through an annulus of area around the float, and out the top of the tube. (See Figure 4-13.) The greater the entering volumetric flow, the larger the annulus required area, and the higher the float rises. The rise is proportional, therefore, to the flow rate. In a simple rotameter the tube body is properly tapered to provide a linear flow scale to reflect the position of the float. Typically the tube is transparent, of glass or plastic, engraved with a scale calibrated in the selected units of flow.

Figure 4-13 Variable area flowmeter

Under the no-flow condition, the float rests at the bottom of the tube. During flow, the forces acting on the float are its weight, buoyant force of the liquid on the float, and the pressure forces acting below and above the float. These forces combine to maintain the pressure drop across the float equal to the effects of gravity and buoyancy acting upon it. For a given fluid, the immersed weight of the float is constant; therefore, at equilibrium the pressure drop or differential pressure across the float is constant. As a consequence, increasing the flow upsets the balance, increases the upward force, and raises the float to provide a wider annulus to pass the flow. The rising float and increasing annulus area allow the fluid velocity and its associated pressure and upward force to decrease until the forces on the float are in equilibrium, and the float stops. When the velocity decreases, the upward velocity pressure follows suit, and the float begins dropping until the decreasing annulus allows the velocity and its resultant pressure to bring the forces back into equilibrium.

The float is fitted with grooves or vanes that cause the flowing liquid to impart rotation to the float. This gyroscopic stabilizing action maintains the float in a central coaxial position with the tube during its up-and-down motion.

Although typically the tapered tube is transparent, in some applications the float is not visible, such as for an opaque fluid or where a metallic tube is used. A metallic tube may be required in cases of large flow volume or flow under high pressure. The float position is then detected by magnetic or electrical techniques to provide an external metering of the flow.

As with other flowmeters, rotameter installations are available which can transmit pneumatic, electronic signals to connect to recorders, totalizers, controllers, and process computers.

Installation

Rotameters are simple to install. They are fabricated in a variety of construction materials to withstand a wide range of pressures and temperatures.

Rotameter manufacturers provide end fitting and connections of various materials and styles to meet customer requirements. Safety-shielded glass tubes are in general use throughout industry. End fittings may be metal or plastic to meet the chemical characteristics of the fluid to be measured. Care must be taken in plastic fitting installations to avoid distortion of the threads.

Plastic tubes are also available in rotameter designs because of their lower cost and high-impact strength.

Liquid measurement rotameters are usually provided with a direct reading scale and calibration data for water. Should the same meter be applied to other fluids, recalibration will be necessary.

Maintenance

Maintenance of rotameters is relatively simple. Where there is potential for dirty flow, an upstream strainer is usually recommended, along with periodic cleaning. In cases where shutdown is prohibitive, bypass pipe may be used to allow flow to continue if the meter is to be removed for maintenance or replacement.

Advantages and Disadvantages

The advantages of variable area flowmeters are

- Low cost
- Ease of installation

- Low maintenance
- Ease of measurement reading
- Linear scale
- Easy detection of faulty operation
- Relatively long measurement range
- Flow straighteners are not required
- No special upstream or downstream conditions are required
- Relatively low and constant pressure drop
- Nearly constant overall pressure loss
- Inherently self-cleaning

The disadvantages of variable area flowmeters are

- Vertical installation required
- Not suitable for dirty fluids
- Not suitable for fluids with suspended solids
- Device is relatively fragile
- Glass metering tubes subject to pressure and temperature limits
- Readings are affected by changes in density or viscosity

OPEN CHANNEL FLOW

When liquid flows in a channel with a free surface, the flow is called open channel flow. Examples include rivers, irrigation canals, and partially full sewers and tunnels not flowing under pressure. To find the flow in an open channel, as in a closed channel, the product of the mean velocity and the cross-sectional area through which it flows would provide the required value. With a known cross-section geometry, the area could be inferred from a depth measurement, provided the mean velocity is easily attained. The mean velocity depends on a velocity profile of the flow stream, requiring multiple measurements, and typically not practiced because other methods are more convenient.

The measurement of flow in open channels is commonly based on hydraulic structures of specially defined geometry. These structures serve as primary flow devices and are built into the flow stream to accommodate the total flow. The major structures in use today are weirs and flumes.

A weir spans the open channel as a dam with a geometrically defined opening over which the liquid is confined to flow. Unlike a weir which partially dams the flow, the flume confines the flow through a specially shaped, open channel structure. In both cases, the geometry of the hydraulic structure and the flow through the device produce a relationship between the flow value and a liquid level measurement taken at a specified location.

A specific, generally nonlinear formula, based on theoretical derivations and verified by test data throughout the range of applications applies to each type of weir or flume. Therefore, open channel flow is calculated on the basis of geometrical dimensions associated with the hydraulic structure type and the measured level or

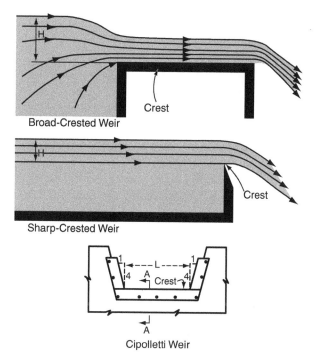

Figure 4-14 Common types of weirs

head. With the hydraulic structure as the primary flow element, the liquid level is determined by a secondary measuring device.

Common liquid level sensors in use are the float, ultrasonic level detector, bubbler tube, and submerged pressure detector. Associated with the sensing of level detection is the conversion to the corresponding flow according to the level-flow formula for the hydraulic structure used. The output measurement may be calibrated to yield level, or the level measurement may be calibrated to yield flow, or the level measurement may be digitally converted to a flow in a microcomputer-based system. The accuracy of the flow measurement depends on both the primary structural device and the secondary instrument device.

Weirs

- Accuracy: ±5 percent of rate to ±10 percent full scale
- Repeatability: ±2 percent
- Rangeability: 75:1
- Size range: 1 ft to 10 ft (0.3 m to 3 m)

Weirs are constructed to confine the flow through a notch, or geometrically defined opening, which may be as wide as the channel. The partial damming effect of the weir structure causes a head or level to develop in a pool of the flow upstream from the weir. (See Figure 4-14.) This depth provides the driving force of flow through the weir and is measured as the level of the pool above the crest of the weir.

The three most common types of weirs are the rectangular weir, the trapezoidal weir (also called Cipolletti), and the triangular or V-notch weir. (See Figure 4-15.) Each type of weir generates a unique formula for flow as a function of the head or

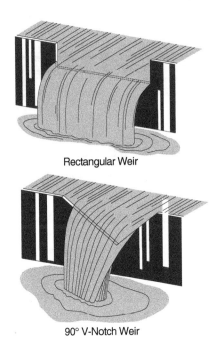

Figure 4-15 Free flow over a weir

depth of the pool measured at a required distance upstream from the weir. The head measurement is the actual level of the pool above the crest (or surface over which the liquid passes) of the rectangular or trapezoidal weir, or above the bottom of the V-notch of the triangular weir. Weirs tend to introduce significant head loss into the stream because the flow must fall freely down from the crest.

Advantages and Disadvantages

The advantages of weirs are

- Simple construction
- Least expensive of primary devices to measure open channel flow

The disadvantages of weirs are

- Higher loss of head than flume (approximately 4 times)
- Deposits collect behind weir, requiring periodic purging.
- Crest must be kept clean.
- Not considered a high-accuracy device
- Excessive approach velocities reduce accuracy.

Flumes

- Accuracy: ±5 percent of rate to ±10 percent full scale
- Repeatability: ±0.5 percent to 2 percent
- Rangeability: 20:1
- Size range: Throat width = 1 in. to 50 ft (.025 m to 15.2 m)

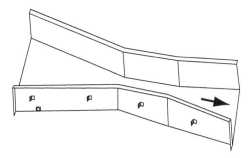

Figure 4-16 Parshall flume

The flume structure comprises a converging section that restricts the flow into a throat or narrowed section, followed by a diverging section where the downstream level may be lower than the level in the converging section. The pattern of deliberate flow restriction and expansion is similar to the effect in a Venturi tube. The head measurement, from which the flow is calculated, is usually taken at a single point in the converging section at a designated distance downstream from the inlet to the flume.

The most popular flume is the Parshall flume (Figure 4-16), developed by Dr. Ralph L. Parshall of the US Soil Conservation Services in 1922. Parshall's modifications in flume design were to downslope the floor of the throat section. This induced higher velocities, tending to keep the flume clear of deposits and maintaining a lower depth of flow.

Advantages and Disadvantages

The advantages of flumes are

- Lower loss of head (approximately ¼) than for weirs
- Minimal effect of approach velocity
- System velocities tend to flush away deposits in the flume
- Dependable measurement with minimum maintenance
- Good repeatability
- Can measure a higher flow rate than similar size weir

The disadvantages of flumes are

- More expensive installation than weir
- Difficult to install

Palmer–Bowlus Flume

This type of flume was developed to meet the need of a primary measurement structure that could be inserted into an existing conduit, typically round, with minimal site requirements and a sufficient slope. The underlying principle of the flume is to provide a sufficient constriction of flow to ensure an increase in kinetic energy; therefore, velocity of flow is greater than the free-flowing conduit into which it is installed. The flume is used most frequently in manholes or open bottom channels to measure flow rates. The flume is typically a temporary installation for gathering flow data to characterize flow at the location so that equipment sizing for permanent installations can be determined.

GENERAL INSTALLATION PRECAUTIONS

The reliability and accuracy of a flowmeter's output signal depend highly on meter installation. An improperly located and installed meter will degrade the inherent specified accuracy below an acceptable level. Responsible manufacturers provide guidance and recommendations for meter installation, which should be rigorously followed. Nevertheless, the meter must be checked immediately after installation and periodically thereafter to assure the stability of calibration and accuracy.

Piping Configurations

Approach piping and, in some cases, discharge piping configurations are important and often neglected installation considerations. Virtually all meters have limitations due to piping configurations, and too often this fact is ignored in piping layouts.

To prevent degeneration of accuracy caused by piping configurations, manufacturers recommend straight pipe lengths both upstream and downstream. The user should acquire manufacturers' tables, which list minimum length of straight pipe required for installation on both sides of the meter.

Swirl, caused by elbow bends or certain pumps in the piping, is a rotary motion of the flow superimposed on the forward motion, which adversely impacts the accuracy of most flowmeters. To prevent swirl, lengths of 50 diameters should be used upstream and 10 diameters should be used downstream.

Fittings

Meters should not be placed in close proximity to a bend, valve, or other fitting that is likely to disturb the flow condition at the meter. Calibration usually takes place under close to ideal conditions with the meter mounted downstream of a long length of straight pipe. If this recommendation is not followed in field installation, the manufacturer's calibration may be invalid. Downstream of the meter, the presence of bends, valves, and fittings may also cause disturbances, which can be transferred back to the meter and affect meter accuracy.

Flow Straighteners

Where installation of the recommended straight pipe length is impossible, a straightener to remove flow distortions, in particular, swirl, should be installed. Care should be taken in using flow straighteners. While they are primarily applied to remove swirl, some may worsen a good flow profile.

Flow straighteners include the tube bundle, perforated plate, etoile, and the Air Moving and Conditioning Association standard (AMCA), as shown in Figure 4-17.

Installation

To avoid causing more harm than good, flow straighteners must be properly installed in the proximity of both the source of flow disturbances and flowmeter itself. In general, straighteners should be inserted at least three pipe diameters downstream of the disturbance source, where some dissipation of the disturbance will have occurred. The distance between the straightener and the meter could ideally be 20 pipe diameters, but not less than 10 diameters, to allow most of the distortion to disappear. In general, the AMCA and etoile flow straighteners produce only negligible head loss. The tube bundle and perforated plate produce substantial head loss.

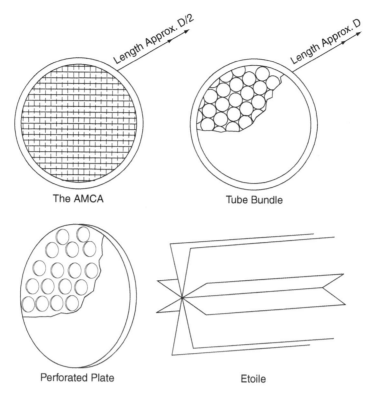

Figure 4-17 Typical flow straighteners

SIGNAL OUTPUT AND TRANSPORT

Signal output and transport are also important for correct meter operation. They are discussed in the following paragraphs.

Flowmeter Output

A flowmeter generates a physical or electrical change that is proportional to flow. If the flowmeter reading is sent to a modern analog panel and digital computer system, the initial physical response must be converted to a signal that has sufficient energy to drive related equipment. This equipment may be an indicator, controller, recorder, or computer.

Because so many applications are available for any sensor output, the instrumentation and control industry has defined a standard: the 4–20 mA DC current loop. In a properly designed loop, the power source and the sensor are connected in a series electrical circuit to the items using the signal. As an example, if these items require a voltage input, where the standard is 1–5 VDC, a 250-ohm resistor in the series loop and across the input to the computer or recorder will provide the proper signal.

Signal Conditioning

The raw initial signals from flowmeters are subjected to signal conditioning, which ultimately generates the standard 4–20 mA DC signal and may also perform many other functions. These include filtering, amplification or attenuation, and linearization. A common signal-conditioning operation for differential pressure flowmeters is to convert the raw signal by a square root function to one directly proportional to flow velocity. The product of velocity and pipe cross-sectional area is commonly computed as part of signal conditioning to generate the desired volume flow.

Signal Enhancement

Microprocessor technology and subminiature electronics are being applied to the conditioning of the signal to enhance the application of flowmeters through programmed functions and menu readouts. Some of these enhancements include:

- Digital rate display
- Digital totalizer display
- Selection of flow spans
- Selection of engineering units in digital display
- Filtering of signal
- Remote location display
- Auto calibration
- Auto diagnostics with alarm readouts
- Coded error messages

In some units a keypad is included to enable the user to select the desired electronic packages. Microprocessors are improving accuracies and lowering costs, while they continue to assume a greater role in the development of all sensors to provide more capabilities and greater flexibility.

REFERENCES

Automation and Instrumentation, AWWA Manual M2. 1983. Denver, Colo.: AWWA.

Bean, H. S. 1971. Fluid Meters: Their Theory and Application. Rept. of ASME Res. Comm. Fluid Meters. New York: Amer. Soc. Mech. Engrs., Sixth Edition.

Cascetta, F. and Vigo, P. 1988. *Flowmeters, A Comprehensive Survey and Guide to Selection*. Research Triangle Park, N.C.: Instrument Society of America.

Cheremisinoff, N. P., and P. N. Cheremisinoff, 1987. *Instrumentation for Process Flow Engineering*. Lancaster, Pa.: Technomic Publishing Company, Inc.

DeCarlo, J. P. 1984. *Fundamentals of Flow Measurement*. Research Triangle Park, N.C.: Instrument Society of America.

Grant, D. M. 1989. *ISCO Open Channel Flow Measurement Handbook*. ISCO, Inc., Third Edition.

Hayward, A. T. J. 1979. *Flowmeters, A Basic Guide and Source Book for Users*. New York: John Wiley and Sons.

Johnson, C. D. 1988. *Process Control Instrumentation Technology*. New York: John Wiley and Sons.

Liptak, B. G. 1982. *Instrument Engineers' Handbook*. Vol. I. *Process Measurement*. Philadelphia, Pa.: Chilton Book Company.

———. 1985. *Instrument Engineers' Handbook*. Vol. II. *Process Control*. Philadelphia, Pa.: Chilton Book Company.

Miller, R. W. 1983. *Flow Measurement Engineering Handbook*. New York: McGraw-Hill Book Company.

Zimmerman, R. and D. Deery. 1977. *Handbook for Turbine Flowmeter Systems*. Phoenix, Ariz.: Flow Technology, Inc.

AWWA MANUAL M2

Chapter 5

Pressure, Level, Temperature, and Other Process Measurements

This chapter discusses the primary sensors associated with three primary process variables encountered in water utility systems: pressure, level, and temperature. The chapter also gives an overview of many less-common sensors used to measure process parameters, such as turbidity, pH, residual chlorine, particles, and stream current. Electric power and equipment status measurements are also discussed.

In this chapter, transducer devices usually refer to those that produce a 4–20 mA DC signal for transmission. Many of these process measurement concepts can also be applied with pneumatic instrumentation where the output signal, proportional to the process variable, is the standard 3–15 psi (20–100 kPa) pneumatic signal. In addition to these output signals, *smart* transmitter devices are available that permit two-way communication from a sensor to a central system. Output signals from process measurement devices discussed in this chapter will normally be used with secondary instrument devices discussed further in chapter 6.

All discussions in this chapter about output signals of various devices refer to analog output signals. Recent developments in sensor technology have made direct digital output signals possible for many of these devices. Digital signal technology is discussed in greater detail in chapter 9.

PRESSURE, LEVEL, AND TEMPERATURE

After flowmetering, three of the most common water system measurements are pressure, level, and temperature. Each is discussed in the following paragraphs.

Pressure

Pressure is an important variable in the operation of water treatment plants and water distribution systems. Pressure readings within the plant permit the operator to observe that pumping systems are operating properly. In distribution systems, pressure readings provide an indication of satisfactory system water pressure to ensure sufficient supply to meet customer demand. Pressure instrumentation may range from simple, direct-reading pressure gauges to complex, pressure-sensing equipment that transmits readings to remote locations.

The simplest pressure device is the direct-reading pressure gauge. In this device, line pressure actuates an element that moves as pressure changes and positions a pointer or recording pen to indicate the measured pressure. The three most common pressure-sensing elements found in pressure gauges are the Bourdon tube, bellows elements, and diaphragm elements. Bellows-type pressure elements are suitable for lower pressure measurements while the others can be used for high-pressure monitoring.

Examples of these types of pressure-sensing systems are shown in Figure 5-1. In each case, the changing pressure in the sensing chamber will move the linkage mechanism and position the indicating element. The linkages can be calibrated so that the gauges read directly in the desired pressure units being sensed. These devices, as shown, will measure pressure above atmospheric pressure, usually referred to as gauge pressure. They can be modified to indicate absolute pressure when required.

Gauge pressure (psig) is the pressure in excess of normal atmospheric pressure. Absolute pressure (psia) is the gauge pressure plus the standard atmospheric pressure of approximately 14.27 psi (98 kPa). Gauge pressure is most frequently used in water system applications. Similar devices can also be used to indicate vacuum or negative pressure.

These devices can also be used as the primary sensing element for a transmitting mechanism by linking a linear, variable, differential transformer (LVDT) to the movable linkage. The armature of the LVDT repositions itself to produce a change in output voltage. The voltage can be amplified to a standard 4–20 mA DC signal for transmission. Figure 5-2 shows an LDVT attached to a Bourdon pressure element.

When the pressure of corrosive fluids needs to be measured, a corrosion-resistant, sealing diaphragm assembly is normally supplied. In this configuration, the pressure-sensing element itself is filled with a liquid from which all air has been evacuated. The noncorrosive liquid is separated from the pressure measurement point by a corrosion-resistant flexible diaphragm forming a sealed capsule but open to the element. The corrosive material presses against this diaphragm changing pressure in the noncorrosive fluid.

An example of a diaphragm seal is shown in Figure 5-3.

Most pressure-transmitting devices use a very low displacement pressure-sensing measurement as opposed to the relatively high volume devices discussed above. Although manufacturers use a number of approaches to design pressure transmitters, the variable capacitance pressure cell is most popular (refer to Figure 5-4). In this device, pressure against an isolating diaphragm is transmitted to a sensing diaphragm through an oil-filled chamber. This pressure is compared against a corresponding diaphragm on the opposite side which is open to atmosphere.

PRESSURE, LEVEL, TEMPERATURE, AND OTHER PROCESS MEASUREMENTS   103

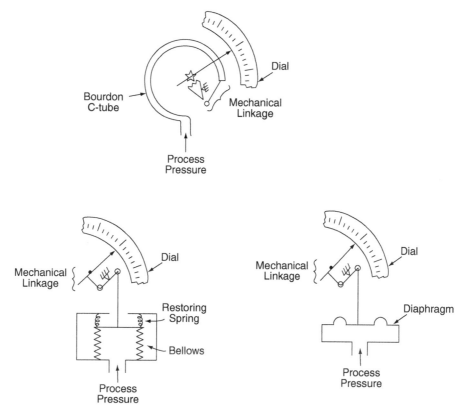

Figure 5-1   Bourdon, bellows, and diaphragm pressure sensors

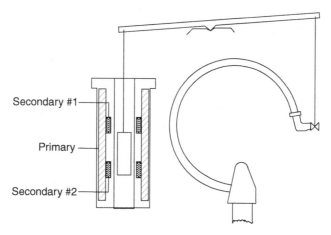

Figure 5-2   Typical LVDT application

The change in pressure in the fluid-filled chamber causes a small displacement of the sensor diaphragm. This change is detected as a change in capacitance between two capacitance plates on either side of the sensing chamber. The change is then converted electronically to a conventional 4–20 mA DC signal.

This type of sensing mechanism can also be used for flow monitoring when used with a differential pressure-producing primary flowmeter, such as a Venturi tube or an orifice plate. In this application, the high- and low-pressure connections of the primary flowmeter device are connected to either side of the sensing chamber. The sensing

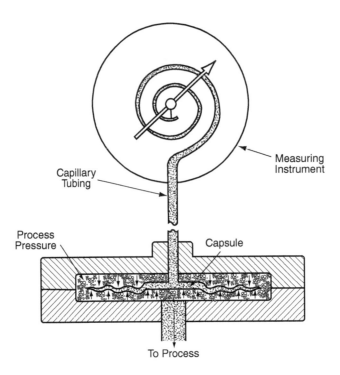

Figure 5-3 Diaphragm seal

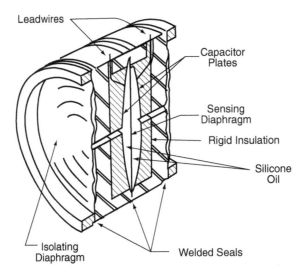

Figure 5-4 Variable capacitance pressure sensor

mechanism then responds to the difference in these two pressures and produces an output signal representative of flow through the primary flowmetering device.

Other concepts used in similar devices include variable reluctance, bonded strain gauges, and vibrating wires. Variable reluctance sensors are similar to the variable capacitor concept except that the electrical quantity is reluctance. In bonded strain gauge sensors, the pressure acts on a semiconductor strain gauge, which causes a change in output voltage. The change corresponds to the pressure acting on the strain gauge assembly. In the vibrating wire concept, pressure changes the tension on a resonant wire. The change in the resonant frequency of the wire is representative of the pressure.

With differential pressure transmitters used for flow measurement, additional electronics in the transmitter circuit extract the square root of the differential pressure measurement. Therefore, the transmitter produces an output signal directly proportional to flow.

Pressure-sensing devices typically have accuracies varying from ±1 percent to less than ±0.25 percent of full-scale measurement. The accuracy depends on the quality of the equipment and care taken in calibration.

Level

Level measurements determine water level in storage tanks or level of liquid chemicals in chemical storage tanks.

Pressure sensors can be readily adapted to level measurement by installing the sensor at the base of a tank. As the level increases in the tank, the pressure reading increases. The reading can be calibrated in feet of liquid. In elevated tanks the level measurement needed is the level in the elevated portion of the tank rather than in the tank and riser. Transmitting mechanisms can be calibrated so that zero represents the bottom or minimum level in the elevated portion of the storage tank.

The float-operated level sensor is another common level-sensing system. Float-operated sensors would normally be used in open reservoirs, particularly when a transmitter cannot be located at the low point of the reservoir. A float-operated transmitter consists of a float resting on the surface of the water. The float usually travels in a stilling well and is attached to a pulley and counterweight with a cable. As the float rises and falls in the stilling well, the pulley system turns and positions an indicating pointer or a transmitting mechanism to indicate water level. An example of a float-type, level-sensing system is shown in Figure 5-5.

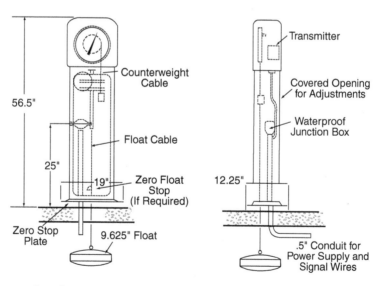

Figure 5-5 Float-type, level-sensing system

106   INSTRUMENTATION AND CONTROL

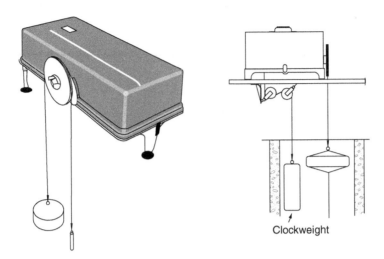

Figure 5-6   Stage recorder

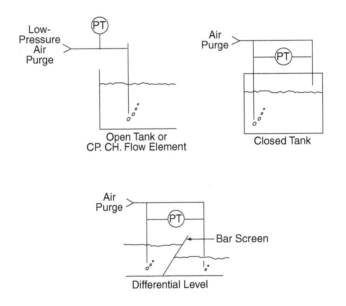

Figure 5-7   Bubbler

Accuracies of conventional float-type level transmitters will normally be ±1 percent of full-scale reading.

Where extreme accuracy in level-sensing systems is required, a special adaptation of a float-type level transmitter has been used. The adaptation is called a stage recorder and uses larger floats and a more precise pulley system to increase positioning accuracy. These devices may also be equipped with a digital sensing system. Levels can be transmitted as binary coded decimal output. Readings are accurate to within ±1/8 in. (3 mm) throughout the full measure range. An example of this type of stage recorder is shown in Figure 5-6.

Another type of level sensor quite frequently seen in water system applications is the pneumatic bubbler (Figure 5-7). A small diameter tube is installed in the tank with the tube bottom as near as possible to the tank bottom. A regulated and

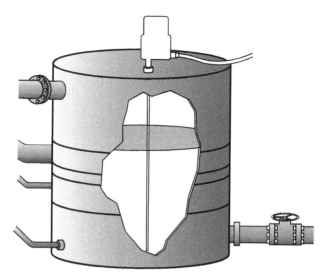

Figure 5-8   Admittance probe

continuous flow of air passes into the tube and slowly bubbles out the bottom end. The air pressure is just enough to overcome the water pressure. A conventional pressure sensor is connected to the bubble tube assembly. The measured air pressure within the bubbler system corresponds to the water depth.

With the bubbler transmitters, a good quality dry air must be used. If the air contains a significant amount of moisture, the bubbler may not operate properly, particularly in freezing temperatures. Accuracies will be ±1 percent of the maximum depth measurement.

In recent years, direct electronic devices have been used for level measurement. Among these are capacitance or admittance probes, variable resistance devices, and ultrasonic systems.

Figure 5-8 is an example of an admittance probe. In this system, an insulated metallic probe is installed in the reservoir. As the water level rises and falls in the reservoir, the capacitance changes between the metallic portion of the probe and the water. This capacitance signal can then be converted to a 4–20 mA DC output representative of level. When the tank is not made of a conducting material, a second electrode will be required. Admittance or capacitance probes can also be used for level measurement of solid or granular material in storage tanks. With some materials, coating may occur on the electrodes that can interfere with proper operation of the admittance probe concept. This possibility needs to be investigated as part of the application of this type of level measurement.

A variable resistance level sensor is shown in Figure 5-9. The sensor consists of a wound resistor inside of a semiflexible envelope. As the level rises, the semiflexible outer portion of the sensor presses against the resistor. A portion of the resistor element is temporarily shorted out, which changes the overall resistance of the sensor itself. The resistance is converted to an output signal representative of level.

In ultrasonic level-sensing systems, an ultrasonic generator is installed above the water level. This generator sends ultrasonic signals toward the water surface. The signals bounce back and are detected by a receiver located with the generator. The time required for this signal to echo is calibrated to produce an output signal representative of water level. An example of an ultrasonic level sensor is shown in Figure 5-10. Air temperature variations must be compensated for because the speed

108 INSTRUMENTATION AND CONTROL

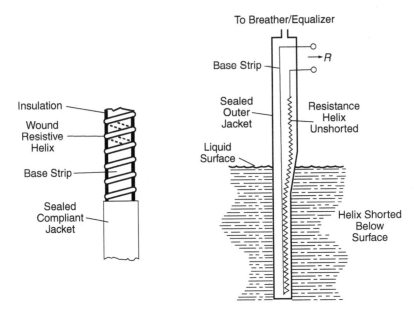

Figure 5-9 Variable resistance level sensor

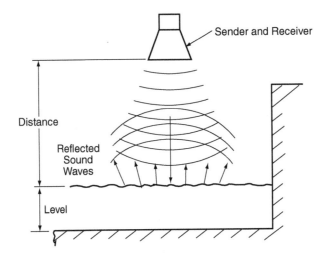

Figure 5-10 Ultrasonic level sensor

of sound in air is a function of temperature. Excessive humidity in the air above the liquid in a stilling well may also significantly interfere with proper operation. Accuracy is ±1 percent of the level measurement.

With the wide selection of level sensors available, the selection of the proper device for a particular application may oftentimes become somewhat confusing. A number of factors must be considered in making this decision. Among others, the factors include the type of material being measured and the physical conditions of the particular installation. Manufacturers' representatives can help in making the choice. Recommendations from those individuals experienced in sensor selection inside or outside the utility are also good sources.

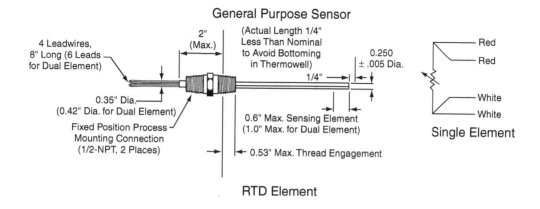

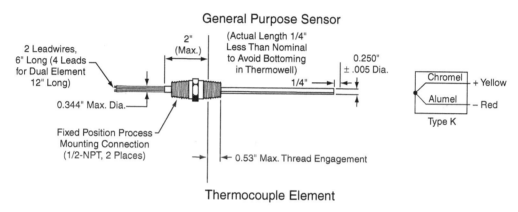

Figure 5-11 Typical temperature elements

Temperature

Temperature sensing is not encountered in water system applications as frequently as level and pressure sensing; however, depending on the specific details of a treatment process, temperature signals may be quite important. Also, temperature measurements often monitor performance of large pumps to detect abnormal operating conditions.

Resistance temperature devices (RTDs) and thermocouples are the two most common temperature sensors. The RTD sensor is a precision-wound metallic element inside a corrosion-resistant sheath. The element is encapsulated in a fill material. Changes in temperature cause a change in the resistance of the metallic element. The change is then calibrated and converted to a conventional output signal for indicating temperature.

A thermocouple is made from the connection of two dissimilar metals. When two thermocouples at different temperatures are connected together, a thermoelectric current proportional to the temperature difference is produced. This low-level signal can be amplified to produce an output signal representative of the temperature difference.

Examples of RTDs and thermocouple temperature sensors are shown in Figure 5-11.

The sensing devices can be installed either directly in the process or device, or in a thermowell. The thermowell is a corrosion-resistant fitting that protects the

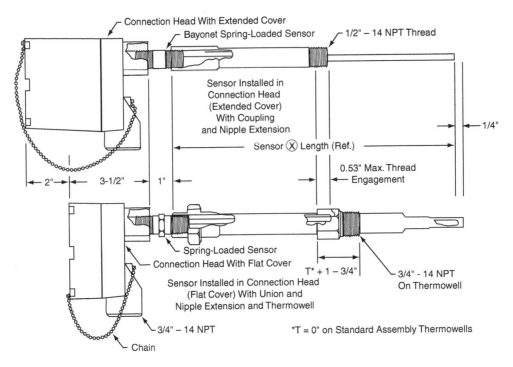

Figure 5-12 Thermowell

temperature-sensing device from direct contact with the process fluid. Figure 5-12 illustrates installation of an RTD- or thermocouple-type temperature sensor with and without a thermowell.

The accuracy of a temperature-sensing system depends on the matching of a sensor and transmitter to meet a particular situation. Generally speaking, sensing accuracies of ±0.5°C can be obtained.

ELECTRIC POWER AND EQUIPMENT STATUS

Not directly related to water or process measurements but equally important are measurements of electric power and equipment status.

Electric Power

With increasing costs for electrical power, the need for monitoring voltage, current, and electric power use is increasing. Expensive, large motors are commonly monitored for voltage and current. Motor voltage and current signals can be converted into 4–20 mA DC process signals for input into the instrumentation system. Figure 5-13 illustrates current transformers that provide output signals proportional to the motor current. A current transformer is mounted on one phase of the motor, and the secondary output of the current transformer is connected to the signal converter. The bottom portion of the illustration in this figure shows a method of scaling that may be useful in matching current transformer output to input requirement of the signal converter. For voltage measurements, similar signal converters are used. The necessary scaling transformer is installed on the input side.

Caution should be observed in providing signal converters for current and voltage monitoring in high-voltage equipment. Particularly in the case of current transformers,

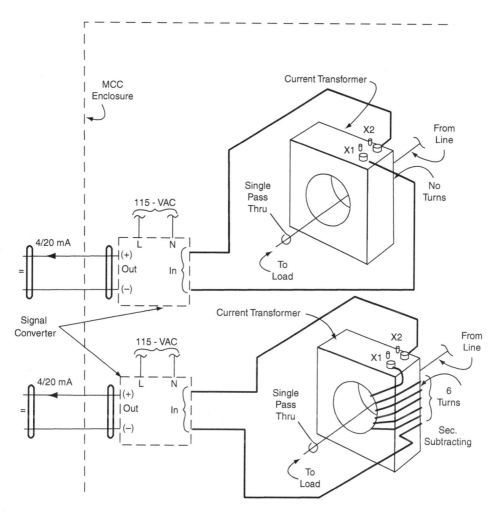

Figure 5-13 Motor current sensor

if the secondary of the transformer is disconnected from the signal converter when current is flowing through the main line, dangerously high voltages can occur on the transformer secondary. For this reason, the signal converters should be mounted within a motor control enclosure. The motor control center must be opened to access the signal converter, and automatic interlocks ensure that power is off.

Current transducers are also available that incorporate an integral transducer to provide a 4–20 mA output without a separate transducer. The transducer portion is loop-powered so that no additional power source is required.

Individual watt transducers can also be provided for individual electric motors to monitor the power being consumed by each device. These transducers monitor both current and voltage and produce an output signal proportional to total power being used by the motor. These devices are suitable for mounting directly in the individual motor control center and incorporate current transducers.

Current and voltage information is primarily used with power and demand monitoring. A more detailed discussion of power monitoring has been included in chapter 2.

Equipment Status Monitoring

A number of operating conditions associated with major equipment should be monitored. Vibration should be monitored in a plant process control system, particularly for large, expensive equipment, such as pumps and blowers. Excessive vibration can quickly cause significant damage to this equipment, particularly when operated at higher speeds (above 1,800 rpm). Vibration sensors are available which, when mechanically attached to the particular piece of equipment, will activate a contact closure when vibration levels exceed a specific acceleration or g value. To minimize problems associated with adapting equipment to accommodate these sensors in the field, vibration sensors could be included as part of the specifications. Analog vibration sensors are also available. These devices produce an analog 4–20 mA DC output signal proportional to the magnitude of the vibration within the specified range of the particular sensor.

Position and speed are two other equipment status conditions that can be monitored. Position transmitters normally work with variable resistance devices mechanically linked to equipment. The output of the variable resistance device is then converted to a 4–20 mA DC signal for monitoring. Speed transmitters are usually tachometers driven by equipment being monitored. The tachometers produce a voltage converted to standard current values for monitoring.

With some equipment, particularly clarifier drives, torque is another variable sometimes monitored. Torque sensing is normally used to shut down circuits to prevent damage to the equipment. Torque-sensing equipment can be supplied to produce either a contact closure when torque rises above a preset value or can be supplied with converters that produce a 4–20 mA DC signal proportional to the magnitude of the torque.

Larger pumps and motors incorporate discrete sensors that measure parameters, such as temperature, pressure, and position. These discrete sensors are installed for safety monitoring and sequential control operations.

PROCESS ANALYZERS

With increasingly stringent requirements for water quality, analytical equipment that measures process variables is used more frequently in water systems. No discussion of primary sensors would be complete without some discussion of process analytical devices that may be used for these applications.

Turbidity

Turbidity monitoring is almost mandatory for the effluent from granular media filters. Raw water turbidity is frequently measured as well as settled water and finished water leaving the treatment plant. These measurements show water quality improvement at different stages in the treatment process.

Early turbidity monitoring equipment used a sensor to detect changes in the amount of light transmitted through a sample. This concept of sensing was found to be nonresponsive, particularly when measuring very low turbidity levels. Currently, turbidity sensing devices monitor the intensity of light scattered from turbidity particles when a beam of light is passed through the sample. The most common method measures the amount of light scattered in a direction at 90° to the path of the light beam. This type of turbidity monitoring has found widespread use for filter effluent turbidity. Figure 5-14 illustrates this principle of light scatter for turbidity sensing.

At higher turbidity levels, this concept will not be practical because the scattered light at high levels of turbidity cannot be measured accurately. One successful method

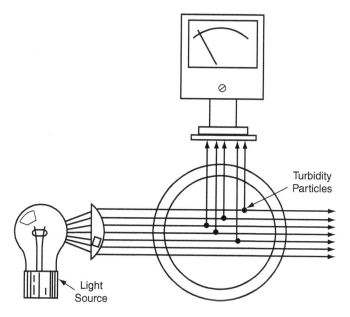

Figure 5-14 Light scatter turbidity

of measuring higher turbidity levels normally encountered in raw water is the surface scatter concept. Light scattered from turbidity particles at the surface of a sample compartment is detected as the sample of water continuously overflows. The surface scatter principle of turbidity monitoring is illustrated in Figure 5-15.

Turbidity monitoring can be most helpful in optimizing plant performance. If proper results are to be obtained, however, a regular program of calibration check is essential with turbidity monitoring equipment. Most manufacturers provide relatively simple calibrating procedures for this equipment.

Accuracy statements with turbidity sensing equipment can be quite confusing because they depend upon the standard used for comparison. Generally, however, accuracies of ±2 percent to 3 percent can be obtained with turbidity monitoring equipment, which is normally satisfactory for treatment process application.

pH

Another analytical instrument used for some time in water treatment plant applications is the pH monitor. pH readings can be extremely useful in monitoring and controlling a treatment process. Chemical reactions affecting coagulation, corrosion, and softening can be predicted, monitored, and changed through proper pH control. pH monitors detect the level of hydrogen ion activity in a sample and convert this to a signal.

A pH sensing system consists of a glass membrane electrode and a reference electrode. The glass membrane electrode develops an electrical potential that varies with the pH of the process fluid. The potential developed between the two electrodes is amplified and converted to a signal representative of the pH. A typical pH electrode installation is shown in Figure 5-16. Details of the electrodes and the instrumentation will vary, depending on the manufacturer.

pH electrodes can incorporate either an immersion-type installation or a flow-through type system. Examples of both systems are shown in Figure 5-17.

Depending on the equipment selected and the calibration techniques, pH sensing systems are accurate in the range of ±0.1 percent to ±0.2 percent pH units.

114 INSTRUMENTATION AND CONTROL

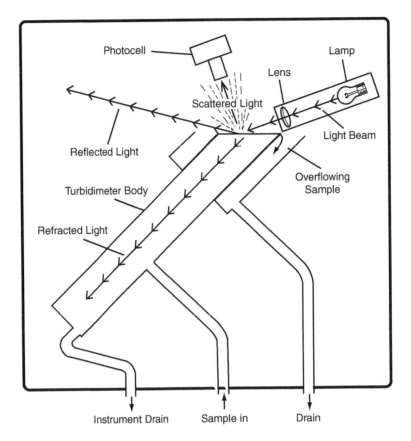

Figure 5-15 Surface scatter

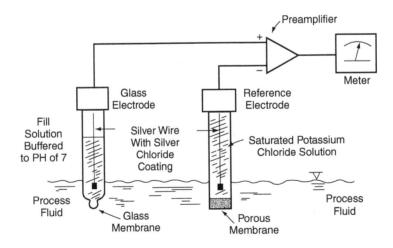

Figure 5-16 pH system

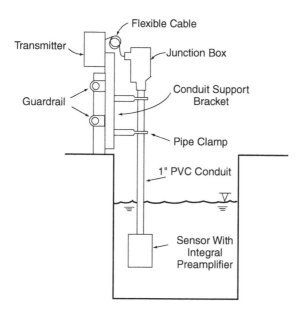

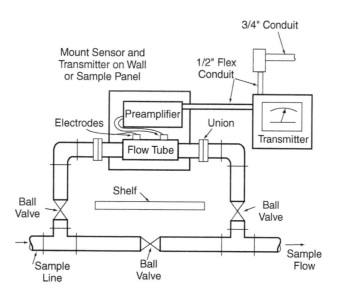

Figure 5-17   Immersion and flow-through pH systems

Residual Chlorine

Residual chlorine monitoring is often required for a treatment process. Two types of residual chlorine monitoring systems are used. One is a chlorine permeable membrane probe which allows chlorine to diffuse through a membrane system on the end of a probe. The chlorine concentration passing through the membrane generates a current in an electrode system that is responsive to the chlorine level. The membrane-type chlorine analyzer probe is shown in Figure 5-18.

A second type of chlorine residual analyzer is the amperometric type. Two dissimilar metals are placed in a measurement cell containing an electrolyte. A

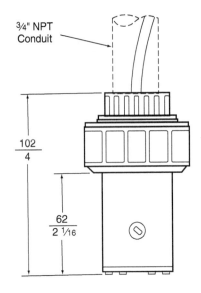

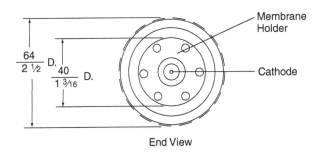

Figure 5-18 Chlorine membrane probe

voltage is applied to these two metal electrodes, and the amount of current flowing between the electrodes is proportional to chlorine present in the solution. A buffer solution and an electrolyte agent are required. A typical amperometric chlorine analyzer configuration is shown in Figure 5-19.

In water treatment applications, total free chlorine is often an important measurement. Because the concentration of chlorine as either HOCl and OCl depends on the pH of the sample, sample conditioning may be required to produce an output signal representative of free chlorine. Normally, if the pH of the sample is 7 or less, sample conditioning will not be required. In most treatment plants, however, corrosion control requirements dictate that finished water pH will be close to 8.2 to 9.0. For correct operation of a residual chlorine monitoring system under these conditions, the water must be buffered to depress the pH signal below 7.0. In the past, liquid buffering solutions have been used; however, today carbon dioxide (CO_2) is often used as the buffering agent. A typical installation of a CO_2 buffering system is shown in Figure 5-20.

Chlorine residual monitoring equipment can normally produce accuracies in the range of ±2 percent to ±3 percent of full scale.

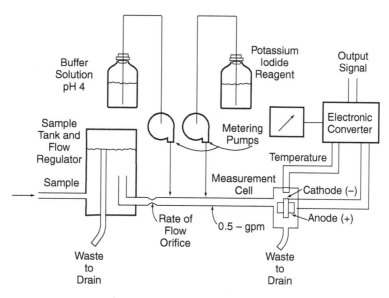

Figure 5-19   Amperometric chlorine residual analyzer

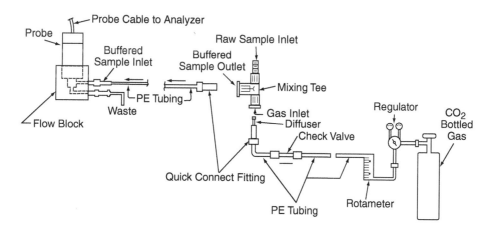

Figure 5-20 CO_2 buffering

Particle Counters

Particle counters are somewhat similar to turbidity monitors; however, they provide a more detailed analysis of suspended matter in the sample. Particle counters transmit a laser beam across a sample stream, which is detected by a sensor on the opposite side of the sample stream. When particles pass through the beam, they interrupt the beam. The sensor also senses particle size based on the amount of the beam that is interrupted. Particle counters can be programmed to provide information on particle count, total particle count below a specified particle size, and particle size distribution. The primary application for these devices has been in monitoring filter effluent in water treatment plants. They also, however, may monitor the effectiveness of coagulation and flocculation.

A typical schematic of a particle counter is shown in Figure 5-21.

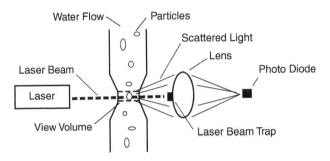

Figure 5-21 Particle counter

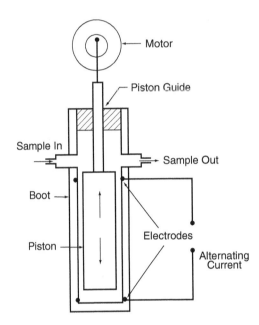

Figure 5-22 Streaming current monitor

Streaming Current Monitors

The streaming current analyzer monitors coagulation in a turbidity-removal water treatment plant. The streaming current monitor measures particle charge remaining after coagulants have been added.

A sample of treated water flows into a sample chamber where it is drawn into the space between a piston and the walls of the sample cell. The piston moves continuously up and down in the sample cell. Sample particles are temporarily immobilized on the piston and cylinder surfaces. Water moves back and forth in the space, which moves the counter ions surrounding these particles. This movement of counter ions generates a current, which is detected by the electrodes in the sample cell. The current is a measured value representative of particle charge. The output from streaming current meters can be used to continuously control coagulant dosage with raw water flow.

A typical schematic of a streaming current monitor sample cell is shown in Figure 5-22.

Miscellaneous Analyzers

Specific ion monitoring systems are similar to pH monitoring equipment; however, they incorporate an electrode specific to the concentration of a substance in the sample solution. Specific ion electrodes are available for an increasing number of parameters, such as fluoride and nitrate. They operate with a reference electrode similar to pH sensors. Properly applied, these devices can be most useful in optimizing treatment plant operation.

Another analytical instrument with wide application in Europe but only infrequently seen in the US is the continuous ultraviolet (UV) spectrophotometer. This analyzer measures the attenuation to a UV light transmitted through a continuous flowing sample. By adjusting the frequency of the UV light beam, these devices can be made specific to a particular element, such as dissolved organic carbon.

Several water quality analytical devices have been designed for monitoring raw water parameters, including conductivity of oil in water. Continuously monitoring raw water quality upstream from the plant intake with these devices will be more widely used in the future.

Another recent development not yet available as a commercial product is direct fiber-optic sensors. Special compounds are bonded to the ends of fiber optics and provide a fluorescent level proportional to a particular component.

GENERAL CONSIDERATIONS

All of the sensor systems discussed in this chapter, of course, depend highly on proper installation to obtain a successful and useful application. The manufacturer's instructions should be carefully followed. A successful installation also depends on proper maintenance and calibration of the sensors.

REFERENCES

American Water Works Association. 1990. *Water Treatment Plant Design*. New York City, NY: McGraw-Hill Publishing Company.

Kawamura, S. 1991. *Integrated Design of Water Treatment Facilities*. New York City, NY: John Wiley and Sons, Inc.

Skreutner, R. G. 1988. *Instrumentation Handbook for Water and Wastewater Treatment Plants*. Lewis Publishers.

This page intentionally blank.

AWWA MANUAL M2

Chapter 6

Secondary Instrumentation

INTRODUCTION

Instrumentation is broadly classified as either primary or secondary instrumentation. A primary instrument is a sensor that measures some process variable, as discussed in chapters 4 and 5. Secondary instrumentation uses these primary instrumentation signals. These instruments are usually panel-mounted and can be in local control panels, filter control consoles, area control panels, or main control panels.

Before World War II, most measuring and control instruments were locally mounted with direct connections to the process. These primary measuring devices required operations personnel to move throughout the plant to take readings and make control adjustments. With a small process plant and an abundant work force, this was sufficient. However, as plant monitoring, maintenance, and compliance requirements increased, monitoring and control functions needed to be more centralized. At first, this concept created a potential safety hazard; when direct measurements were made of processes containing highly toxic material, the control room operator could be exposed to the toxic material should a signal line break or leak. For these reasons, secondary instrumentation was developed. Secondary instrumentation required the development of a signal transmission method as well as field and panel-mounted hardware to perform the monitoring and control functions.

Secondary instrumentation, together with the primary measuring elements (chapters 4 and 5), telemetry (chapter 7), and final control elements (chapter 8) permit the operator to continuously supervise an entire plant's operation from one location.

SIGNAL STANDARDIZATION

The first transmission methods used air pressure, which is a form of pneumatic signal transmission. Later, the use of electrical current or voltage signals—electronic signal transmission—was developed. Both pneumatic and electronic signal transmission perform the same functions, and the instrument exteriors are nearly identical in the field and in the control room. As development and usage have progressed, the signal levels have become standardized. The pneumatic standard is 3 to 15 psi (20–100 kPa), and the electronic standard is 4 to 20 mA DC. The practical advantage of this standardization is that the operating mechanisms of the control room instruments are based on the common signal units in all locations. The panel may contain indicators, recorders, and controllers handling pressure, temperatures, flows, and other process variables, but their input and output signals operate over the same standard range. The only differences between them are the scales or charts used to display the data.

Another advantage of the standardized signal levels of secondary instrumentation is the *live zero*, which makes it possible to determine the variance below a minimum process measurement. As all the instruments operate over their standard 3 to 15 psi or 4 to 20 mA DC range, any reading registering below their minimum values, 3 psi and 4 mA DC, respectively, would indicate the existence of a problem with either the signal transmission system or the transmitting device.

Pneumatic and electronic signal transmission can be used within the same process control system, allowing a great deal of flexibility in designing a treatment facility. Equipment can be chosen that most effectively suits the application and environment. Pneumatic pressure to electric current (P/I) and electric current to pneumatic (I/P) signal converters are readily available.

The basic characteristics of pneumatic and electronic systems are summarized in Table 6-1. The table lists the advantages and disadvantages between pneumatic and electronic systems, both in performance and capability. Because of their advantages, electronic instruments are used for most new installations today, although pneumatic instruments are preferred in hazardous areas and high-temperature applications.

The recent use of microprocessors in primary instruments has made it possible to communicate directly from a computer to the primary elements using data links, such as RS422 or RS485 routers. The reason for using these data links is that the 4 to 20 mA DC signal is only able to communicate one variable in one direction, is susceptible to noise, is less accurate than a digital signal, and is only capable of having its rangeability set at the field device. Microprocessors and data links allow remote calibration of instruments and will eventually eliminate secondary instruments. Data links are sometimes referred to as field buses because the computer data bus is extended into the field. The communications support, data acquisition, supervisory control, and the operator station functions for this type of a distributed control system are discussed in chapter 10.

SIGNAL POWER AND TRANSMISSION

The backbone of any secondary instrumentation measurement is the energy-generation and transmission system used to relay the signals. With the pneumatic system, the energy-generating facility is composed of an air compressor, air dryer, and storage tank. Pneumatic secondary instruments rely on the slight movement of a flapper nozzle mechanism to sense signal change. Therefore, clean, dry air must be provided to ensure proper operation. Essential steps in providing high-quality instrument air are drying, automatic draining of condensation, pressure reduction, and expansion of the final distributed air. Figure 6-1 shows a typical single compressor for a small system. Normally, large systems have two compressors to provide system backup and to permit lead-lag operation for varying air demand. The air supply is typically supplied throughout the plant in ½ in. (12 mm) pipes and

Table 6-1 Comparison of electronic and pneumatic systems

Feature	Pneumatic	Electronic
Transmission distance	Limited to a few hundred feet	Practically unlimited
Standard signal	3–15 psi, practically universal	4–20 mA DC, practically universal
Compatibility between instruments supplied by different manufacturers	Typically not a problem	Occasionally nonstandard signals may require special consideration and may not be compatible
Compatibility with digital computer or data logger	P/I converters for all inputs	Easily arranged with minimum added equipment
Reliability	High if energized with clean dry air	High if provided with proper surge protection
Effects of temperature extremes	Poor in low-temp applications	Poor in high-temp applications
	Superior in high-temp applications	Superior in low-temp applications
Reaction to electrical interference (pickup)	No reaction possible	Potential for EMF interference
Operation in hazardous locations (explosive atmospheres)	Completely safe	Intrinsically safe
Reaction to sudden failure of energy supply	Superior: capacity of system provides safety margin, backup inexpensive	Inferior: electrical failure may disrupt plant, backup expensive, battery backup available
Cost of installation	High	Low
System compatibility	Fair: requires considerable auxiliary equipment	Good: conditioning and auxiliary equipment more compatible to systems approach
Cost of maintenance	Fair: procedures more readily mastered by people with minimum of training	Good: depends on training and capability of personnel; equipment must be removed for most maintenance
Dynamic response	Slow, but adequate for most situations	Fast: frequently valve becomes limiting factor
Operation in corrosive atmospheres	Naturally unaffected, air supply becomes a purge for most instruments	Requires special protection to be added
Measurement of all process variables	No	Yes

40 to 60 psi. This pressure is then reduced to a standard 20 psi for operation of each pneumatic instrument. Pneumatic signals are typically transmitted through tubing, usually ¼ in. (6 mm) copper or plastic, from field to control room and within the control room between instruments.

Power supplies for electronic systems are available in many sizes and types. One is shown in Figure 6-2. The type most commonly found in instrumentation systems are AC-to-DC power supplies. In these devices, the input voltage is usually 120 VAC, and the output voltage is usually 24 VDC. The size of a power supply is stated in amperes of output current at 24 VDC. Power supplies can also have

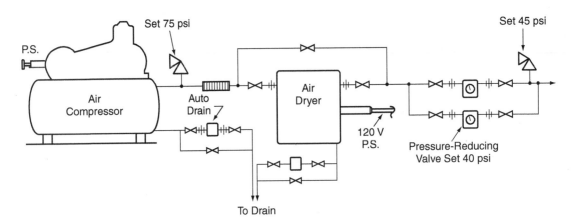

Figure 6-1 Typical single compressor system

overvoltage and overcurrent protection and can be redundant or provided with battery backup for critical applications.

Electronic field instrumentation is classed as either a two- or four-wire device. A four-wire device has two wires for power and two wires for the analog signal. A magnetic flowmeter is an example of a four-wire device. Two-wire devices receive excitation power via the analog current signal wires. This power is usually provided by a panel-mounted controller or multiloop power supply mounted in the local control panel. Two-wire devices are used wherever possible to reduce installation costs because they do not require a separate electrical supply.

TRANSMITTERS

A transmitter internally converts a sensor's output into a standard signal that can be remotely transmitted to panel-mounted hardware for display, recording, and control action. All pneumatic instruments are variations of the same basic mechanism comprised of a sensor (usually a diaphragm or bellows), a flapper-nozzle assembly to detect errors, a bellows to provide feedback, and a relay to amplify the output signal and to re-balance the system.

Two-wire electronic transmitters typically use bellows or diaphragms in conjunction with linear voltage-differential transformers, capacitance plates, or strain gauges to detect errors or changes in the process variable. The resulting error is rectified and amplified to produce a new output signal proportional to the measured variable. Other components, such as span and zero adjustments, are typically provided.

Four-wire transmitters typically measure analytical properties, such as pH, turbidity, and chlorine residual. These devices employ electrochemical sensors which produce small voltage or current signals based on the measured variable. These small signals are then amplified to produce the standard 4 to 20 mA DC outputs.

CONTROLLERS

Controllers provide automatic or manual control of the process, allowing the operator to adjust the final control elements to match the process requirements. In the automatic mode, the controller senses the process variable input value and generates an error signal (difference between input and set point). The controller then

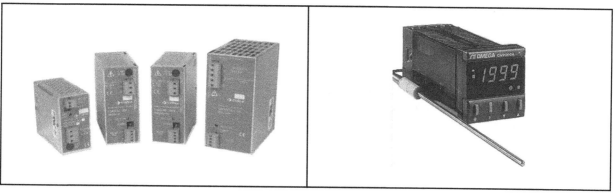

Figure 6-2 Power supply   Figure 6-3 Basic controller

generates an output signal to the final control element so that the error is reduced. The basic controllers shown in Figure 6-3 provide the following interfaces for the operator:

1. A set-point adjustment that produces a signal to be fed into the controller mechanism.

2. The manual control unit for switching between manual and automatic modes of operation and directly manipulates the controller output when in manual.

3. Indicators to display the measured value and set-point signal.

4. An indicator to display the controller output, typically in percent of full scale.

The basic controller provides proportional, integral, and derivative (PID) automatic feedback control functions:

- Proportional—the output of the controller is continuous and linearly proportional to the input signal

- Integral (reset)—the output signal is continuous and proportional to the time integral of the input error

- Derivative (rate)—the output is continuous and proportional to the rate of change of the input error

PID control actions are normally combined to give the best control response of the process being controlled. The normal combinations are: proportional, proportional-plus-integral, proportional-plus-derivative, and proportional-plus-integral-plus-derivative. Additional information, examples, and uses of PID control can be found in chapter 9.

Basic controllers allow for configuration of the following features:

- Set-point limiting

- Output limiting

- Alarming

- Feedforward (flow pacing)

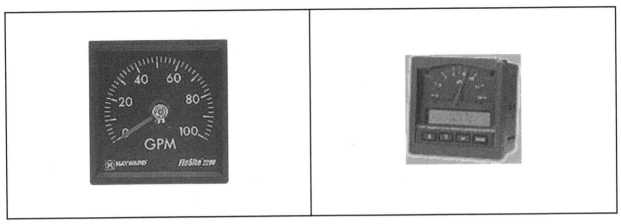

Figure 6-4   Analog indicator

Figure 6-5   Analog and digital indicator

- Preset manual output
- Reverse/direct action
- Bumpless, balanceless transfer between automatic and manual control

The first electronic controllers were constructed of discrete electronic components (transistors) and duplicated the PID control actions of pneumatic controllers. These first electronic controllers were all single-loop controllers, and manufacturers built them in different versions ranging from manual loading stations to computer-manual-automatic-tracking stations.

Microprocessor-based controllers contain the basic PID algorithm, but they also have increased calculation and data storage capacity, allowing a greater number of features and more powerful control strategies. Microprocessor controllers are built as single controllers with all of the functions of previous generation electronic controllers preprogrammed into them along with additional enhanced function modules. The user simply selects control functions from a menu.

RECORDING AND INDICATING HARDWARE

All recorders and indicators receive a standard signal generated by the transmitter and convert it to pen or pointer position. They provide the operator with a constant visual display of input or output variables, allow the operator to double-check automatic controls, and support manual control of the process when required.

Indicators can be either analog or digital. A typical analog meter is shown in Figure 6-4. The digital indicators are typically either light-emitting diode or liquid crystal display. Some indicators have both analog and digital displays, as shown in Figure 6-5.

An indicator can identify the process measurement value fairly accurately on a scale with its pointer, which can be seen from a distance. A recorder tracks the process measurement value continuously through a pen's tracings on a chart and must be observed at close range. Some instruments provide the best of both and have an indicator scale as well as a recording chart; some controllers have recorders built into them.

The chart or scale specifies the process measurement range and units, not the standard signals range and units, and the two should not be confused. For example,

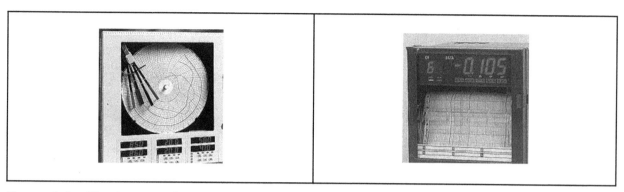

Figure 6-6 Circular recorder

Figure 6-7 Strip chart recorder

in a recorder with a 3 to 15 psi (20 to 100 kPa) input signal and a chart range of 0 to 100 psi (0 to 700 kPa), the process measurement value of 0 psi would be represented by a 3 psi (20 kPa) signal, a midrange process measurement value of 50 psi (350 kPa) by a 9 psi (60 kPa) signal; and a full-range process measurement value of 100 psi (700 kPa) by a 15 psi (100 kPa) signal.

Recorders usually track up to four variables on one chart through the use of separate pens, each with a different color ink. Each recorder has a chart drive, moving the chart under the pens at a specific speed. Although all variables combined in one recorder must use the same time scale, different recorders can have different time scales.

Recorder charts come in many variations. Depending on which of two primary types they are, charts are either circular, covering a certain time period such as a week, or log strips of paper on rolls or fan-folded like computer paper. Charts have different process scales and time scales printed on them. The appropriate type for any recorder is chosen on the basis of how fast the chart is traveling, how much of it should be easily accessible, the readability desired, the space availability, any possible governmental requirements, and the frequency of maintenance desired. Generally, some trade-off of these features is required.

The recorded time scale has to do with how fast the chart travels; it can vary from $3/4$ in./min to $3/4$ in./hr (19 mm/min to 19 mm/hr) for strip charts, and from one rotation per 24 hours to one rotation per 7 days for circular charts. Generally, the faster the process changes, the faster the chart should move to save and record those changes. A circular chart, which occupies more control panel space than a strip chart, can simultaneously display more data than a strip chart, but the circular chart holds less historical information.

Recorders in water treatment plants are used primarily to record both influent and effluent flows. In larger plants, recorders are also used to record backwash flow, turbidity, chemical usage, and chlorine residual. These recorders can be single- or multiple-pen recorders. The recorders can be circular, as shown in Figure 6-6, or strip charts, as shown in Figure 6-7.

Another type of indicating device is a totalizer, which is available in two basic types depending on the input variable. Totalizing counters add up pulse-type signals from flowmeters, such as turbine or propeller meters, or magnetic flowmeters. Their readings indicate the total volume of fluid that has passed through the flowmeter. The integrating totalizer is used to integrate flow rate signals rather than pulses. Integrating totalizers perform the mathematical function of integration.

128 INSTRUMENTATION AND CONTROL

FUNCTION MODULES

A function module is a computing device that accepts one or more standard input signals, modifies or combines the signals in some predetermined mathematical way, and transmits the resultant output to a receiver (e.g., an indicator, totalizer, or controller). Function modules can be mounted in the field or in a control room. They are economical and ideal for a relatively small or uncomplicated system. Some of the common types are described in the following paragraphs.

Integrator. The integrator computing module integrates a flow rate signal over time and provides an output proportional to the current sum. Integrators frequently incorporate a square-root extractor function because the most common input signal is the output of a differential pressure measuring device, which produces a signal that is the squared value of a flow rate. When built into a counter, these devices are referred to as integrating totalizers.

Multiply/divide. The multiply/divide module accepts one to three standard input signals, modifies or combines them according to a mathematical function (multiplication, division, squaring, or square-root extraction), and provides a standard output signal proportional to the result of the calculation. Each unit can perform all four functions and can be changed in the field when control schemes are altered. Applications include mass flow computing, stock blending, flow proportioning, and others used in feedforward and feedback control systems.

Sum or differential. The summation module typically receives between two and four standard input signals and combines them, using addition or subtraction to produce a standard output.

Enhanced function modules. Most of the enhanced function modules were developed and implemented in the first generation of distributed control systems in which eight or sixteen single-loop controllers were implemented by a single microprocessor mounted in a panel interior. Now manufacturers have packaged up to four single-loop controllers into the same 3 in. × 6 in. (75 mm × 150 mm) enclosure for face panel mounting.

These enhanced controller function modules can be configured by using a built-in keypad, separate handheld configurator, personal computer, or downloaded from a distributed control system workstation, if so connected.

The typical enhanced function modules can perform any of the following functions:

- Compensated gas flow control
- Single-station cascade control
- Dual input override control
- Error squared controller
- Manual/automatic station
- Ratio/bias
- External reset controller
- Sampling or flow pacing control
- Self-tuning
- Addition
- Subtraction

- Multiplication
- Division
- Square root
- Compare
- Log (base 10 or *e*)
- Exponentiation
- Swap (*X* <–> *Y*)
- Duplicate
- Absolute value
- Totalize
- Analog input/output
- AND
- OR
- XOR
- Invert
- Duplicate
- Digital input/output (data link)

These enhanced features allow both analog and digital control to be combined. Both continuous analog and sequence batch control can be implemented by a single enhanced controller.

CONVERTERS

Converters are used to convert signal levels or types so that a wider group of components can be incorporated into an instrument system. P/I and I/P converters are used to interface between the two standard secondary instrumentation systems.

Some converters are used as amplifying devices to convert millivolt and milliamp field signals to standard output signals. Examples of this are resistance temperature devices and thermocouples. Another application of signal converters is the current to current (I/I) transmitter usually receiving a 4–20 mA DC input signal and copying this input into another 4–20 mA DC output signal. This converter is useful to isolate the groups of two-instrument systems. It is also useful for powering additional instruments in a complicated loop. For example, if the maximum loop resistance the field instrument can drive is exceeded by the other instruments in the loop, an I/I transmitter can be added to make the loop work correctly.

This page intentionally blank.

AWWA MANUAL M2

Chapter **7**

Telemetry

As water facilities became more widely distributed, data and control signals had to be transmitted over longer distances than secondary instrumentation could provide. Telemetry systems were developed to provide long-distance communication.

Any telemetry system consists of three main types of components: a telemetry transmitter, a receiving unit, and the telemetry channel or communication channel that connects the transmitting and receiving devices. A simplification of this is shown in Figure 7-1.

In most systems, these three main components are technically known as the RTU (remote terminal unit), the SCADA system (supervisory control and data acquisition), and the communication protocol and physical communication network. RTUs are the remote devices that collect data, typically from field devices, and transmit the data. RTUs may also perform local control functions. RTUs acquire data through electrical signals connected to the RTU or from other devices via a serial data connection.

A SCADA system consists of one or more computers, often in a network. (More details on SCADA systems, computers, and networks are found in chapter 10.) The system provides an interface to the RTUs through the communication network. The SCADA system stores the data, displays them, analyzes them, and transfers them to other computers if necessary. A SCADA system often provides a control interface for sending data or commands to RTUs.

A communication protocol is the language used in the transmitting and receiving of data messages. Both the transmitter and receiver of the data message must use the same protocol so that both understand the data. A protocol usually includes the following information:

- the identification of the device sending the data
- the device receiving the data
- the meaning of the data
- the verification information that ensures the complete message arrives and that it is error free

132 INSTRUMENTATION AND CONTROL

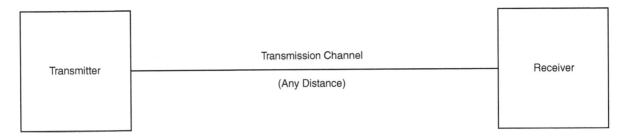

Figure 7-1 Telemetering

The communication network allows the transfer of data from an RTU to a SCADA system, from an RTU to another RTU, and in some architectures between multiple SCADA systems. Many communication network technologies are used with telemetry, and the design of the communication network is critical to the effective operation of a telemetry system.

Several types of telemetry systems are available and in use by water utilities. The system types may be classified into two broad categories: analog and digital. In analog telemetry, the measured variables such as flow, level, and pressure are converted into the value of the telemetered signal. In digital telemetry, the measured variable is sensed initially as an analog electric signal, then converted to a digital message consisting of a series of coded pulses, which, when decoded by the receiver, represents the magnitude of the measured variable.

The major difference between analog and digital telemetry is that in analog telemetry the signal is continuously transmitted. In digital telemetry, the signal does not have to be transmitted continuously but is transmitted as digital data representing a particular instantaneous value of the process variable. The analog signal is displayed as the position of a pointer on an indicator of the position of a pen on a recording chart. A digital signal, however, is represented by a numerical display of the absolute finite value.

Both analog and digital telemetry can be further subdivided into different specific types of systems. Under the broad category of analog telemetry are pulse duration modulation (PDM) telemetry, pulse frequency or pulse count telemetry, variable frequency telemetry, and variable current and voltage telemetry. In digital telemetry, the two major concepts are unidirectional and bidirectional. Unidirectional involves a remote station continuously reporting to a central control center through digital messages. Bidirectional means that the control center addresses the remote station and requests data transmission. These systems normally operate in a polling mode with the central unit sequentially addressing all of the remote stations.

In analog telemetry systems, the transmitter measures the process variable and converts the measurement into a signal. The signal travels to the receiver through a transmission channel. The telemetry receiver converts the signal into a mechanical position to actuate a pen in a recorder or a pointer in an indicating instrument.

In digital systems, the remote or transmitting device is normally the RTU, and the receiver is known as the control terminal unit (CTU). A significant difference between analog and digital telemetry systems is that in digital systems, the RTU does not itself directly measure the variables in the system. Instead, the variables are measured by a transducer that converts the value of a process variable into a current output signal of 4–20 mA DC. These signals are transmitted to the RTU, which converts them into data that are transmitted to the CTU. A typical diagram of a digital telemetry circuit is indicated in Figure 7-2.

In water distribution networks, analog telemetry signals will normally be transmitted over leased telephone lines or privately owned wireline communication circuits. Digital telemetry can be transmitted over a wider variety of communication circuits, such as leased telephone lines, private wirelines, radio, microwave, and satellite communication links. Communication media and channels are discussed in more detail later in this chapter.

Current and voltage signals do not normally apply to remote analog telemetry applications because they do not work when the distance exceeds one mile. From a practical standpoint, the types of analog telemetry normally used are pulse duration modulation telemetry, pulse frequency telemetry, and variable frequency telemetry. These main types are discussed in more detail in this chapter.

Most telemetry systems installed today are digital. Because computers are used in all phases of utility operation, having all data available in a database offers many advantages over analog systems which display status information on indicating and recording receivers. However, many analog telemetry systems still operate, especially in smaller applications.

In a large telemetry system, combinations of analog and digital applications can be economical. As an example, if only a single bit of data must be sensed, such as a distribution system's pressure in a remote location, an analog telemetry subsystem may be more economical. The single analog value could be transmitted to the nearest digital RTU, where it can be added to the database of that RTU for transmission to the central control location.

ANALOG TELEMETRY

The most common analog telemetry systems are discussed in detail in the following paragraphs.

Pulse Duration Telemetry

The most popular analog telemetering method is PDM. In PDM, the signal generated for transmission by the remote transmitting device is a pulse having duration proportional to a quantity being measured. The magnitude of the signal is the length of this electrical pulse, and normal variations in the actual electrical magnitude of the pulse have no effect on the accuracy of signal transmission. As long as sufficient power is on the circuit for a pulse to be received, the transmitted signal will be correct.

A simplified schematic of a typical PDM telemetry system is shown in Figure 7-3. In this representation, a pressure transmitter is used to transmit a PDM signal to a remote telemetry receiver. In the pressure transmitter, the pressure measuring element, in this case a Bourdon tube, positions a pointer with relation to a cam that is driven by a constant speed motor. For simplicity, in this diagram, the transmitter switch is shown at the end of this pointer. The switch is closed when the pointer or cam follower is on the cam and opened when the pointer is off the cam. At zero pressure the cam pointer will be at the outer edge of the cam; at maximum pressure it will be near the center of the cam. The dotted circles on the illustration represent these two positions.

As the cam rotates, the wiper will be on the cam for a shorter period of time at zero or low pressure and a longer period of time when at maximum pressure positions. Under these conditions, the transmitter switch will remain closed for a longer interval of time out of the total cam cycle as the pressure increases, producing

134   INSTRUMENTATION AND CONTROL

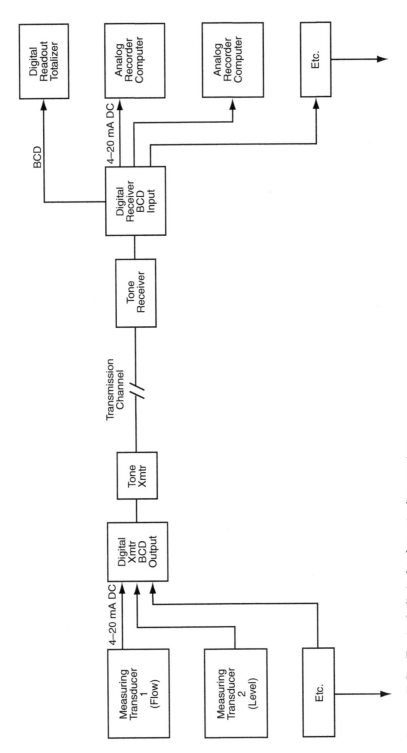

Figure 7-2   Typical digital telemetering system

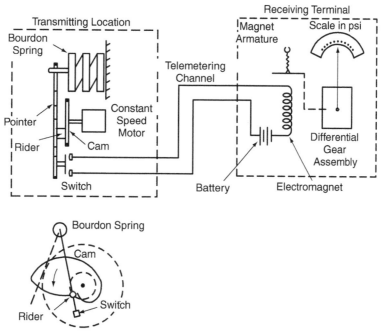

Figure 7-3 Schematic of a typical PDM system

a longer pulse. In this manner, the transmitter generates a pulse, the duration of which is proportional to the magnitude of the measured pressure. The total transmitter cycle, that is, the sum of the time the switch is open and closed, will always total the time required for one rotation of the cam.

A communication channel of some type links the transmitter to the receiver, where a suitable power supply is connected in a series with an electromagnet. When the transmitter switch is closed, the electromagnet at the receiver is energized, and its armature is engaged. When the transmitter switch opens, the electromagnet is de-energized, allowing the armature to be pulled away by some type of spring. The relative motion of the armature acts through a differential gear assembly, which in turn positions a pointer on a scale marked in units of the measured variable. In this manner, the receiving mechanism converts the length of the transmitted pulse into a proportional position of the pointer on the indicating scale, which provides an indication of the magnitude of the remote measured pressure signal. Solid-state-type receivers are also available for use with pulse duration telemetry signals. These receivers incorporate a servo-type pen-positioning mechanism, which compares a feedback potentiometer on the output of the pen-positioning mechanism to an input voltage signal developed by converting the proportional pulse duration signal length to voltage. This type of receiving mechanism has the advantage of eliminating continuously moving electromechanical-type devices.

By controlling the profile of the cam in the transmitter mechanism, the relationship of the cam follower position to the length of the transmitted pulse can be varied to solve simple mathematical equations. This concept is very useful when transmitting flow values based on nonlinear measured variables, such as differential pressure measurements from flowmeters, flow over weirs, or through Parshall flumes. The cam can be used to create a linear signal.

PDM telemetry is relatively slow because of the time required for the continuously rotating cam. The cam makes one complete revolution in a specific period of time depending on the speed of the motor. Although different cam cycles have been used, most PDM telemetry in use today uses 15-second cam cycles. For this reason, it can be seen that normally at least 2 cam cycles would be required to ensure that a correct value has been transmitted. Speed is not an important factor in most water distribution system applications because measured variables do not change very rapidly.

Solid-state approaches to pulse duration telemetry are also available. In these systems, the continuously rotating cam is replaced by a solid-state transmitter configuration. This includes an electronic transducer, which is used to convert the process variable, such as pressure or flow, into a 4–20 mA DC signal. This signal is converted into a proportional switch action in a continuous 15-second cycle, which duplicates the signal characteristic developed by the rotating cam type of electromechanical transmitter. This signal can then be transmitted over the same communication circuit to a receiver mechanism.

Pulse Frequency Telemetry

Pulse frequency telemetry systems are used when the measured variable can be easily converted into a series of on–off pulses. The rate of on–off pulses is proportional to the value of the measured variable. One typical application of pulse frequency type transmitters is a propeller flowmeter equipped with a transmitting switch that is actuated by the rotating shaft of the register mechanism. The flow of water through the pipe drives a propeller or turbine at a speed proportional to the velocity of the flowing liquid. The propeller rotates proportionally to the flow rate. The on–off signals are generated by coupling the shaft of the propeller to a rotary electric switch. The number of contact closures in a given unit of time becomes proportional to the flow rate. The gearing in the meter head is normally used for calibration purposes to provide a specific number of pulses for each quantity of flow through the meter. Typically, these meters will have full-scale pulse rates of 0–100 pulses per second for 100 percent meter output.

The signals from a pulse frequency transmitting device are then transmitted to the receiving unit. A pulsed frequency telemetry receiver can be anything from a simple totalizing counter, which counts up the pulses as they are received, to an indicating–recording receiver. In the pulse frequency receiver, a receiving mechanism responds to these pulses, positioning an indicator or recording pen in proportion to the pulse rate.

Variable Frequency Telemetry

When fast response in analog telemetry is essential, variable frequency telemetry equipment will often be the most satisfactory. The delays in signal transmission experienced in pulse duration telemetry do not occur in variable frequency telemetry. One typical application of variable frequency telemetry is the monitoring of a rapidly changing system pressure where delays would inhibit monitoring the instantaneous peak values.

In variable frequency telemetering, the measured variable is normally sensed with a transducer providing a current output signal of 4–20 mA DC. This signal is transmitted to a variable frequency transmitter. At the receiving end, the variable frequency incoming signal is converted back into a 4–20 mA DC signal. This signal is transmitted to an indicating or recording receiver that indicates the measured variable in engineering units. Typically, the frequency varies from 10 to 30 Hz for a change in measured variable from 0 to 100 percent. Other frequency ranges may be used depending on characteristics of the particular devices. With variable frequency telemetry, audio tone transmission will normally be used for signal transmission.

The transmission channel in earlier telemetry systems would normally be a two-wire metallic electric circuit with DC current being applied to the signal circuit by the transmitter switch. This normally permitted a single signal to be transmitted over one pair of wires. Generally, simple two-wire metallic circuits can no longer be obtained from commercial telephone companies. Modern telephone systems use sophisticated, digital, central office systems that provide communication circuits for voice-grade communication. Direct current signals can no longer be transmitted through leased telephone circuits. Audio tone telemetry equipment can be used with these types of circuits, and it also provides the opportunity for frequency multiplexing of multiple signals over a single voice-grade telephone circuit. Audio signals can be handled quite efficiently by the telephone systems with amplification as required to ensure proper transmission of the signal. Tone multiplexing makes it possible to transmit up to 20 separate variables over a single communications circuit.

TONE MULTIPLEXING

Tone multiplexing can be used with variable frequency, pulse count, and pulse duration telemetry when multiple signals are transmitted over a single communication channel. Tone multiplexing may also be used in digital telemetry.

The two basic types of tone multiplexing are amplitude modulation and frequency shift keying. Audio tone transmitters and receivers operate on specific frequencies within the range limits of 420 to 3,000 Hz. These frequencies normally apply to standard voice-grade communication circuits. Higher frequencies are available where higher-quality transmission lines are provided, as with conditioned telephone lines and microwave transmission channels.

Each audio tone transmitter and receiver pair operating on a given transmission line is referred to as a channel. Each transmitter generates an audio tone at a specific frequency and occupies a specific portion of the frequency spectrum. This audio tone can be detected on the communication circuit when a headset is attached to the communication line. Normally, audio tone channels are spaced at 100 Hz intervals. The characteristics of the communication channel limit the keying speed of the tone. Higher keying speeds will tend to interfere and intermodulate between channels and increase transmission losses. Where faster keying speeds must be transmitted, a higher quality circuit providing a wider bandwidth must be made available to avoid interference. The speed at which information may be transmitted with audio tone multiplexing directly depends on the bandwidth provided by the communication channel. The total number of tones of channels that may be used on a single transmission line depends on the spacing between adjacent tone channels. As with conventional analog telemetry, 100 Hz spacing is normally used. For high-speed transmission, greater channel spacings of 120 to 600 Hz are available. These higher channels are normally used as carriers in digital telemetry systems.

AMPLITUDE MODULATION TONE

Amplitude modulation (AM) tone channels generate a single frequency that is keyed on and off the transmission line. In other words, the presence or absence of tone on the line indicates whether the signal is on or off. The corresponding AM tone receiver is equipped with a selective filter on the input that is tuned to the specific frequency of its corresponding transmitter. The tone receiver provides an output signal when its particular tone is received and ceases to provide the output signal when that frequency is no longer on the line. Tone receiver output signals can either be voltage outputs or relay contact outputs. AM-type tone channels are susceptible to noise and interference that may be present on the transmission line and cause improper response of the tone receiver. AM tone is seldom used today; however, operating installations of AM tone equipment will occasionally be seen.

FREQUENCY SHIFT KEYING TONE

Frequency shift keying (FSK) is the tone equipment used in most modern systems. FSK provides a much greater degree of reliability and flexibility compared with AM tone equipment. FSK tone equipment has an inherent advantage of significantly greater noise rejection than AM tone equipment. With FSK tone equipment, the tone transmitter generates a center or carrier frequency that is constantly on the transmission line. When a signal is transmitted, a variation in frequency or a frequency shift of this carrier frequency permits transmission of the data signal.

When the FSK transmitter shifts the frequency to a higher output frequency called *mark* and then to a lower output frequency called *space*, this provides three-state operation. This operation can be used for simple remote control operation of valves and motors. FSK tone channels are also used in a two-state operation to transmit pulse duration type analog telemetry. The actual shift in frequency is normally ±25 Hz for the 100 Hz spacing. For higher spacing, the frequency shift will be higher. Because the carrier frequency is always being transmitted, its absence can serve as a line failure alarm when the corresponding FSK tone receiver is equipped with carrier detect output relay.

COMMUNICATION MEDIA AND CHANNELS

Telemetry data and control signals travel through communication media such as wire, or through wireless channels such as radio waves. Choosing the right media depends on the signal type and use. The main types of media and channels are discussed in the following sections.

Copper Wiring

Copper wire has been used for many years to convey analog and status information. Cable systems and their associated connectors have well-established standards covering their applications. Traditionally, they are low cost and are readily available from multiple sources. Service personnel can install and maintain a system with minimal training. The principal limitations of copper wire are its capacitance and its inductance. Both of these factors limit a particular wire to a definable distance at a specific frequency.

Twisted-pair and coaxial cable are common copper wires in digital communication. Twisted-pair wires, which consist of wires twisted together to reduce noise, are usually surrounded by a shield to further reduce noise levels. Shielding is used to provide a barrier to electrostatic, electromagnetic, or induced coupled paths. Shielding increases the distance capabilities of a copper wire system. Twisted-pair communications are restricted to relatively short distances and low frequencies.

Coaxial cables are used where higher data transmission rates and longer distances are required. Coaxial cables commonly have a single core or wire surrounded by a thick insulating barrier. A metallic shield and outer environmental cover is placed over the insulation. Prior to fiber optics, coaxial cables were the preferred media for local area network (LAN) communication. Coaxial cable systems are typically higher in cost than twisted-pair systems and require more complex electronics and specialized connectors.

Telephone Lines

Telephone systems in digital communications use either leased lines or dial-up service. Leased or dedicated lines are special lines provided by the local telephone service, which continuously connect the remote site with the main computer. These wires are routed through the switching offices of the phone system. In situations where the remote site lies in the area covered by a different phone network, several switching centers may be involved. Traditionally, many systems used leased lines to economically monitor remote sites. Since deregulation of the telephone system, the cost of many of these services has greatly increased. Often failures of telephone company systems make leased lines inoperative.

A second method of remote telephone service is a dial-up network. Either the computer or associated circuitry dials the phone number of the remote site. Once the connection to the remote modem is made and *handshakes* validated, data are then transmitted. This is very similar to using fax machines. Dial-up systems are the least expensive method of remote communications but also the slowest. The requirements for the majority of remote sites, however, do not require rapid data updates. Transmission rates from 300 to 19.2K bits per second are available in many locations. Furthermore, dial-up systems do not suffer the same failure rate as leased lines.

The vast majority of telephone lines in service were designed to handle voice-grade communications. Using these lines to transmit digital information limits the rate at which data can be sent. Depending on location, several telephone circuits may be available. Many telephone systems offer specialized digital communication circuits for computer usage such as integrated services digital networks and digital subscriber lines. The advent of fiber-optic technology has also opened new network services and possibilities.

Fiber Optics

Fiber optics use light to transmit data over a fiber cable. Fiber-optic cables have become popular in LANs; see chapter 10 for more information) because of their noise immunity. Traditional coaxial cables are subject to noise from a wide variety of electrical devices. Although theoretically a faster media, current fiber systems in LANs do not offer a substantial improvement over copper systems; however, fiber systems do operate at high speeds over longer distances than copper systems. The cost of fiber cables is becoming more competitive with copper systems.

Radio Systems

Radio telemetry systems offer an alternative to fiber lines, wire lines, or telephone lines for the purpose of communications with remote sites. The US Federal Communication Commission (FCC) regulates radio frequencies and their applications. License applications may take three to six months to process and must be considered in the startup of a project. The FCC will usually only grant one frequency per system, so it must be shared among all the remote sites.

A typical radio system consists of a master station, main antenna, remote antennas, remote radio/modems, RTUs, and repeaters, if needed. Repeaters are used where distance or terrain do not allow the master station proper communication with the remote site.

Radio systems work in a similar manner to dial-up telephone service. The remote sites are commonly polled in sequence, and data are transmitted once the connection is verified.

The major considerations in any radio system are

- Distance to the remote locations
- Terrain between the master and remote sites
- Frequency used
- Output power of the transmitter
- Sensitivity of the receiver
- Height of the antenna
- Type of antenna

A listing of nomenclature for frequencies is shown in Figure 7-4.

Radio systems are terrain-sensitive. Radio communications commonly require a line of sight between the transmitter and receiver. Hills, trees, buildings, and other obstructions may degrade or prohibit reception. The frequency selected will affect the system design. Very high-frequency signals are better able to bend around hills than are ultrahigh frequency (UHF) signals but are subject to greater problems with natural and anthropogenic noise and co-channel interference. UHF and microwave frequencies require line of sight and higher antennas. A radio signal's ability to penetrate solid walls decreases with an increase in frequency. However, a radio signal's ability to enter through smaller and smaller openings increases with an increase in frequency. The appearance of improved penetration comes from the increased ability of UHF signals to enter buildings through doors, windows, and other small openings and the improved ability to bounce around once inside. These higher frequencies, however, suffer from severe attenuation from anything in their transmission path. Trees and foliage can severely reduce system capabilities. The frequency used will also affect cost. Normally, the higher the frequency, the higher the cost of the associated equipment.

Frequency Subdivision	Frequency Range
VLF (very low frequency)	Below 30 kHz
LF (low frequency)	30 to 300 kHz
MF (medium frequency)	300 to 3,000 kHz
HF (high frequency)	3 to 30 MHz
VHF (very high frequency)	30 to 300 MHz
UHF (ultrahigh frequency)	300 to 3,000 MHz
SHF (super-high frequency	3 to 30 GHz
EHF (extremely high frequency)	300 to 3,000 GHz

Figure 7-4 Nomenclature of frequencies

The demand for radios in some areas has reduced or eliminated the availability of many frequencies. Heavily industrialized cities have limited frequencies available. As with the telephone system, most radio usage is directed to voice transmission. Data-quality radios are, in many cases, more costly than their voice-grade counterparts.

Trunking Systems

As stated previously, because of demands in some areas, frequency availability is becoming a problem. Several manufacturers are offering a proprietary, common radio system that combines the usage of many of the municipal services. In a trunking system, police, fire, and other city services are serviced by a central radio system. Each shares time using selected frequencies. The system is prioritized to allow the most critical request through first (police and fire). In normal operation, the scanning speed of remote sites is provided adequate coverage. Trunking systems are primarily voice-grade systems. Digital data radios are provided at a premium.

Spread Spectrum Radio

Spread spectrum radio is a new technology for data transmission that is more widely used for SCADA applications in water systems. Spread spectrum technology uses low power 900 MHz radios that do not require FCC license approval. This technique involves frequency hopping over an available 240 separate narrow channels to locate an available clear channel. Communication from the master to remote sites is accomplished by *skipping* from one site to another until the target site is reached. Poll cycles and data collection times may be somewhat longer than is the case with dedicated 900 MHz radio systems. In areas where dedicated frequencies are no longer available, spread spectrum technology may offer an option.

Satellite Links

In some isolated applications, satellite telemetry provides a radio link at a high cost. But these systems are not limited by terrain or distance. These systems are common in oil and gas transmission lines where the industry can bear the added cost.

Cable TV

Some cities use local cable TV hookups as mechanisms for data transmission. These provide high-quality, full-time paths for system operation similar to leased telephone lines. Currently, this option has several drawbacks. Interface equipment suitable for SCADA applications are not commonly available. Also, tariff regulations must be applied, and most cable networks do not have a mechanism in place for billing this type of service.

Hybrid System

In small, single-purpose applications, the system may require only one communication scheme. In most municipal applications, both in-plant control and remote monitoring are needed. Large systems may link multiple plants as well as provide the management databases for operations. Interfacing with wastewater operations is also becoming necessary with many water utilities. The compatibility of various telemetry systems is a function of the operating software. The selection of the software should be based on its ease of interface with a variety of remote telemetry as well as local operation. Software designed for industrial plant automation usually does not support the needs of remote telemetry.

Future Developments

Telecommunications and data transmission is a rapidly changing field. New technologies are always under development. In developing the communication strategy for any new SCADA system, all possible communication options should be examined.

REFERENCE

Aubin, P. 1997. *Telemetry Systems, Techniques and Standards*. Instrumentation and Control Exhibition Symposium, June 1997. <http://www.iica.org.au/letters/telem.html>.

AWWA MANUAL M2

Chapter **8**

Final Control Elements

An effective control system includes a sensor (primary instrument) to observe and transmit the current state; a controller (secondary instrument) to determine if the current state is acceptable, and if not, what adjustment is necessary; and a mechanism (final control element) to alter the process. In a water system, two major elements can produce a change in the process: valves (or gates) and pumps. Pumps increase pressure and flow while valves (or gates) decrease pressure and flow.

Final control elements (FCEs) consist of signal conditioners, actuators, and final elements, as shown in Figure 8-1.

Signal conditioners. Signal conditioners receive either pneumatic, electric, or electronic signals from a controller. A signal is then conditioned or amplified, and used to control the actuator. Signal conditioners include diverter valves, starters, and positioners.

Actuators. Acuators produce movement, either rotary or linear, of the final element. They require external energy, such as electricity, high pressure air, or hydraulic pressure, and are usually motors or cylinders and their related gearing.

Final elements. Final elements are pumps and valves that change the process fluid.

Understanding the relationship between these three elements is important. In many cases, the elements are supplied by different vendors and each component must be carefully selected to achieve the overall system requirements. Each combination of these three will produce an FCE with its own distinct characteristics. As shown in Figure 8-1, the FCE is an *integral* part of the total control system.

Two types of control are required in a process control system: two-state or modulating (also called continuous). Two-state control requires the FCE to be either on or off, or opened or closed, etc. Modulating control requires the FCE to be continuously adjustable so that it can operate at any point between a preset minimum and maximum value.

In addition to the two types of control and the three components of final control elements, a variety of primary forcing medias (air, electricity, hydraulic oil, or water) and several actuator types are available. This chapter introduces the most common types of FCEs found in water systems but not all the possible combinations and applications.

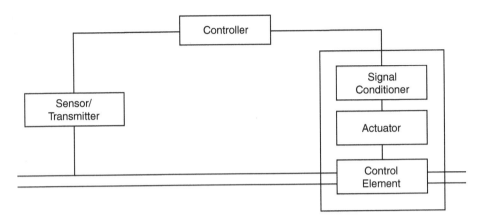

Figure 8-1 Components of control

VALVES

The three elements of valves as FCEs are discussed in this section. Signal conditioners are presented first.

Signal Conditioners

Signal conditioners can be two-state or continuous.

Two-state. Two-state conditioners are either diverter valves for use with pneumatic and hydraulic actuators or electric switching circuits for use with electric actuators.

Diverter valves. In pneumatic or hydraulic operated actuators, a *diverter valve* (four-way or three-way) applies directional force to the actuator cylinder. The diverter valve will normally have two positions. In one state, the primary forcing media (air, water, or oil) is ported through the diverter valve to one side of the actuator cylinder. The other side is either ported to the atmosphere or a low-pressure return line to the supply system. The pressure in the primary media forces the actuator to move in one direction until it reaches its end of travel. In this position, the diverter valve maintains the force in the chosen direction. When the diverter valve moves to its other position, the primary forcing media reverses to the opposite port of the cylinder, forcing the actuator in the other direction (Figure 8-2). The diverting valve movement is most often performed by energizing the solenoid coil (Figure 8-3). A spring or second coil returns the valve to its original position. Proper sizing for the application is critical. In some special applications, air or fluids may move the diverting valve. An alternative configuration that is often used to accomplish two-state control uses two three-way diverter valves. This configuration permits stopping the valve at an intermediate position.

Electric switching circuits. An on–off service with an electric actuator uses switching circuitry as shown in Figure 8-4. In small systems, a simple switch directly energizes the motor windings, causing movement in one direction or the other. The system will stay energized until it reaches the end of travel where the limit switch will shut off the motor. The gearing of the actuator will then maintain the valve position. In larger electric actuators, the control signal will be applied to a reversing starter which in turn will energize the windings of a three-phase motor.

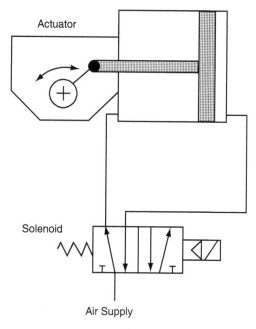

Figure 8-2   Solenoid with cylinder actuator

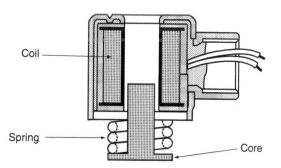

Figure 8-3   Solenoid with details

Modulating service. Modulating service signal conditioners are called positioners and can be used with pneumatic and hydraulic or electric actuators.

Pneumatic and hydraulic. In the on–off application, only two positions are required. In modulating (continuous) services, an infinite number of positions are required. The input to the positioner is normally either 3–15 psi (20–100 kPa) for pneumatic control systems or 4–20 mA for electrical control systems. Current to pneumatic converters are often used to change a 4–20 mA signal to 3–15 psi (20–100 kPa) air.

A pneumatic positioner uses a specialized four-way valve to achieve continuous operation. The resultant valve position is proportional to the control air signal input. The typical arrangement has 3 psi (20 kPa) as closed, 15 psi (100 kPa) as open, and 9 psi (60 kPa) as midpoint.

As the control air input increases (see Figure 8-5), it pushes the positioning valve against the position feedback spring until the positioning valve ports (directs) air to the actuator cylinder, which moves the process valve further open. The rotating

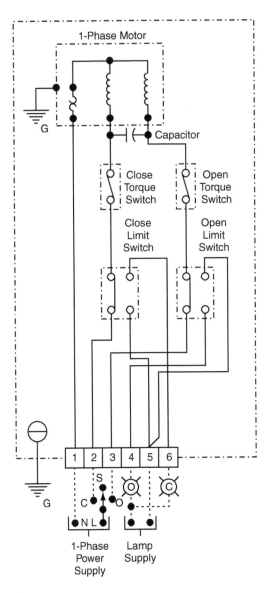

Figure 8-4 Single-phase motor

action of the cam increases the balancing force of the spring, causing the positioning valve to move back to the mid-position. The mid-position of the positioning valve blocks the air pressure from the actuator cylinder, stopping the valve at the new position, which is primary forcing proportional to the new control signal valve.

The quality and quantity of the primary forcing media (air, oil, water, or other fluid) must be controlled. Most system problems can be traced to contaminated forcing media. Control system maintenance should include proper filtration and conditioning of the actuating medium.

Electric. Early electric valve positioners used costly direct current (DC) motors. Newer alternating current systems use either single- or three-phase motors to produce positioning (Figure 8-6). The actuator position is controlled by an external analog command signal. The servo amplifier accepts inputs of current, voltage, or resistive command signals. The solid-state comparator circuit compares the command

FINAL CONTROL ELEMENTS 147

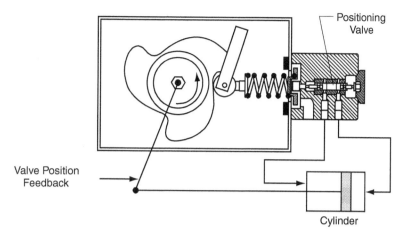

Figure 8-5 Pneumatic positioner cut away

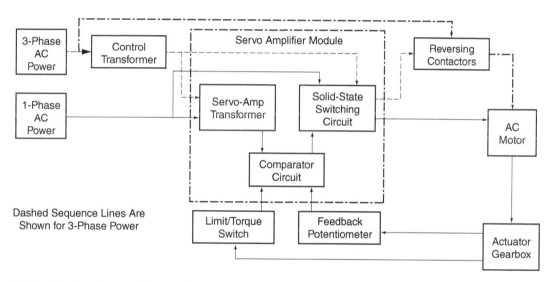

Figure 8-6 Electronic positioner circuitry

signal with the position of the actuator via a feedback potentiometer. The switching circuit incorporates open and closed power triacs. Power is supplied through the triacs to the motor until the required actuator position is obtained. An override circuit is provided through travel limit and torque switches. When the circuit is open, either at the end of travel or in an over-torque condition, the electric power to the actuator motor will be removed.

Traditional electronic or pneumatic control systems have limitations inherent with their design. The quality of the analog input signal can be easily degraded by factors such as temperature changes, distances, and adjacent wiring. As computer control becomes more acceptable, direct digital communications are being used. Binary coded digital information can be directly passed to a microprocessor-based positioner and used to initiate valve movement. Digital control is discussed further in chapter 10.

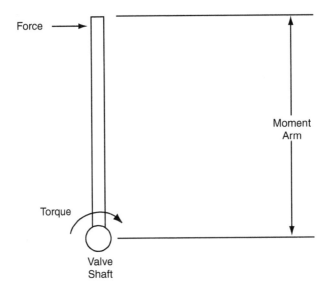

Figure 8-7a Rotary valve requires torque

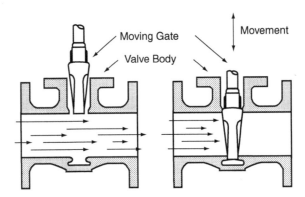

Figure 8-7b Linear valve requires thrust

Valve Actuators

Actuators provide force to move the valve to a required location. This force can be either rotational (torque), as with butterfly valves, or linear (thrust), as with gate valves. Valve movement occurs when the actuator can overcome the frictional and dynamic forces of the valve. Figures 8-7a and 8-7b illustrate the two forces.

The required force is not a constant and is a function of the valve type and the dynamic forces of the fluid on the valve. For most valves the greatest amount of force is required to move the valve into and out of its seated or closed position. As the plug (disc, ball, or gate) leaves the seat, the required force decreases dramatically. In some applications, the dynamic forces of the fluid are not always in a single direction. Depending on valve design, this force can either assist or resist the valve movement. In some valves, the dynamics of the fluid may produce a closing force for part of travel and an opening force for the remainder of travel.

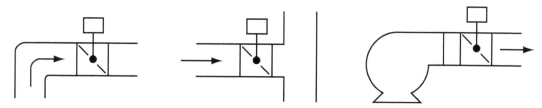

Figure 8-8   Piping configurations

Proper actuator selection is based on system pressure and flow as well as valve size, orientation, flow direction, and the upstream and downstream piping configuration (Figure 8-8).

Valve actuators consist of three subcomponents: a power mechanism, a linkage, and a connector. The power mechanism is a motor or cylinder that converts electrical, pneumatic, or hydraulic energy into rotational or linear force. The linkage, or gearing, typically uses mechanical principles to multiply the force of the power mechanism. A connector joins the actuator to the valve. The connector must be adequate to transmit all required forces and tight enough to eliminate any mechanical slippage. In modulating service applications, all connections must be solid and connected in a manner which will tolerate rapid and continuous mechanical reversals.

Valve actuators are categorized by their power requirements, either electric, pneumatic, or hydraulic. Pneumatic units use air, and hydraulic units use water or oil as a power transmitting media. The choice as to which is the most suitable requires an analysis of the entire system. Each actuator type has its strengths and limitations.

Electric actuators. The power supplied to an electric actuator is usually single-phase or three-phase. The choice of power requirements is a function of actuator force required and power available. With on–off service, the actuator motor performance characteristics should match the valve needs. For example, at the time of its highest load the motor is generating its maximum power, and therefore its highest running temperature. As the load drops off, the energy required from the motor decreases and the temperature drops. In a continuous control system, the motor is expected to be able to start and stop repeatedly anywhere throughout the range of valve travel. As a result, it is designed to operate at higher temperatures than on–off applications.

If electric power is cut off for any reason, the mechanical connection of the gearing, if properly sized, will resist any of the dynamic forces from the valve to move. Also, electric actuators usually come with manual override systems, which allow the unit to be operated by hand.

Pneumatic and hydraulic actuators. In pneumatic and hydraulic actuators, pressure is supplied by a compressor, pump, or plant water system. The actuator in pneumatic and hydraulic systems most commonly uses a cylinder attached by linkage to the valve. The fluid is diverted into one side of a cylinder and the increased pressure moves the cylinder in the other direction. Diaphragm actuators are also used but cost and size restrict their usage to small valves.

The most common fluids used in pneumatic or hydraulic actuators for water systems are air and water. Systems using plant water as the supply media are becoming less common. High-pressure oil actuators are considered where remote failsafe systems are needed. One important issue in these systems is the cost of operation and preventive maintenance for the compressors and pumps required to supply the pressure.

Valve Selection

Proper valve selection requires a complete understanding of the properties of the media being controlled, the type of control needed, and the capabilities and limitations of the different valve types. American Water Works Association has standards for a variety of different valves, although these standards do not cover all valve types and are further limited to specific fluids, pressure flows, and temperatures. The major components are the body, a movable component (disc, plug, or gate) shaft or stem, and a seat.

The body carries the fluid and provides the outside boundaries. The body must also provide the mating surfaces for the adjacent piping. In rotational valves (butterfly, ball, plug, cone) the disc, ball, plug, or cone is connected to a shaft. In linear valves (gate and globe style), the gate or plug is connected to a stem. Lifting the stem or rotating the shaft moves the internal member through its range of travel.

The design of the seating mechanism has changed more than any other valve component. Originally, seats were metal-to-metal contact, which are prone to leakage. Rubber compounds and early elastomer provided better closure but had many limitations. Today, the advances in the chemical industry have produced materials that are both compatible with the fluid and can provide bubble-tight shutoff at high pressure and flows.

As a valve moves through its entire range of operation, its effect on the process fluid varies. This relationship is defined as the valve characteristics, and each valve has its own characteristics. Two general categories of characteristics are *inherent* and *installed*. Inherent characteristics are a function of the specific valve components as manufactured and are quantified from test results or theoretically determined using standard operating conditions. Installed characteristics are based on the actual performance in a particular system. Adjacent piping and flow path all affect installation characteristics, and rarely are the installed and inherent characteristics equal.

Three main valve classes are quick opening, linear, and equal percentage (Figure 8-9). Quick opening valves produce large changes as the valve initially opens. The rate of change decreases greatly over the later portion of valve travel. Equal percentage characteristics are opposite of quick opening. Large flow rate changes occur as the valve nears the open position. Linear characteristics occur where flow rate changes correspond directly to valve position movement.

The signal conditioner, the actuator, and the valve all have their own distinct characteristics curves. To properly operate a control valve each part must be considered in its relationship to the combined assembly.

As the valve opens, the fluid rapidly passes through the valve. High velocity may be detrimental to the seat; metal and elastomer components can be eroded or damaged. Operating a valve at or near its full closed position is undesirable. A valve should be operated over the range of greatest linearity.

Valve sizing is a method of applying the correct size valve to a particular control situation. Mathematically, valve sizing attempts to use the most linear portion of a valve characteristics curve. Valve sizing equations address flow, pressure drop across the valve, and the *CV* (flow coefficient) for a specific valve. Standard valve sizing formulas are provided by many sources and are beyond the scope of this manual.

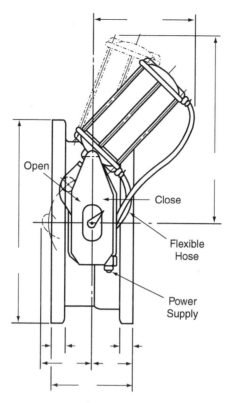

Figure 8-9 Control characteristics

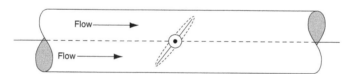

Figure 8-10 Butterfly valve

The following paragraphs briefly introduce the major types of valves.

Butterfly valves. Butterfly valves are rotational valves. The disc of a butterfly valve is always located in the flow stream. Therefore, these valves are not suitable when unobstructed passage is required. At full closure, the circular valve disc is positioned 90° to the flow stream. As the butterfly valve rotates open, the angle of the disc changes with relation to the body. In the full open position, the disc is parallel to the flow (Figure 8-10). Butterfly valves used in water systems are typically greater than 3 in. (75 mm) in diameter.

Ball valves. Ball valves are rotational valves. The rotating ball can either be full ported to match the pipe diameter or reduce ported. In the full open position unobstructed passage through the valve is provided. As the ball rotates, the passage reduces and flow direction through the valve changes. At the full closed position the port is perpendicular to the flow path. A recessed seat provides tight closure. Control at low-flow rates near the open position produces a sluggish response. Ball valves usually require larger actuators because of their high torque requirement. Ball valves are usually used in small chemical feed lines or in large pump-check applications where full ported valves are required.

Cone valves. Cone valves have many of the same operating characteristics and similar applications as ball valves. Using a metal seating surface, the rotation of the cone requires a three-step process. The tapered cone is first lifted a small distance inside the valve body to break the metal-to-metal surface contact. Then the cone is rotated to the desired position. The tapered cone is then dropped or pressed back into the valve body, reseating the metal surfaces.

Plug valves. Plug valves are rotational valves and are concentric or eccentric. These valves use waxes and liquid lubricants as sealing devices. Concentric plug valves, similar in appearance to cone valves, are used primary in open–close gas operation.

Eccentric plug valves appear as modified cone or ball valves. In ball and cone valves, the closed position results in two surfaces being placed perpendicular to the flow; however, only one is used in seating the valve. The design of the eccentric plug valve uses only half of the ball or cone valve.

In the full open position, the plug is out of the flow stream recessed into the valve body. The eccentric rotation of the plug into the seat produces a camming action forcing the plug tight against the seating surface. These valves are used in sludge and slurry applications (Figure 8-11).

Gate valves. Gate valves are linear valves. Gate valves come with a rising stem or a nonrising stem. Using the principle of the screw, the disc is raised or lowered by rotating the shaft. Gate valves are normally not used in continuous control applications. Linear movement of the shaft produces wear and reduces valve life. Gate valves are typically used in distribution systems (Figure 8-12).

Globe valves. Globe valves (sometimes called *angle* or *y-pattern* valves) are linear valves. Control is achieved by raising and lowering the plug with reference to the seat. A variety of plugs can be used to produce different control characteristics. This style of valve does not provide for a straight flow path. Globe valves are used in applications where pressure control is needed. Globe valves are normally used to control remote tank levels (altitude valves) and excess pressure (surge control) (Figure 8-13).

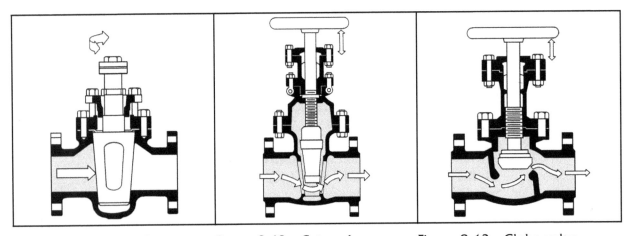

Figure 8-11 Plug valve   Figure 8-12 Gate valve   Figure 8-13 Globe valve

Sleeve valves. High-pressure applications require special valves to reduce and control pressure. One common valve used is the sleeve valve. The internal sleeve is raised or lowered into the main body. The sleeve contains numerous ports for the fluid. Movement of the sleeve controls the number of ports exposed, controlling the flow through the valve.

Other Valve Considerations

Besides the three main elements of valves, other considerations including cavitation, feedback, and failsafe systems are important.

Cavitation. Cavitation occurs inside a piping system when a vapor *implodes* to a liquid caused by an increase in fluid pressure. A sudden drop in fluid pressure to a point less than the vapor pressure causes liquid to flash into vapor. This drop can occur, for instance, as the fluid experiences increased velocity and decreased pressure as it passes through a valve. As the fluid exits the valve, it experiences decreased velocity and increased pressure. If the increased pressure is above the fluid's vapor pressure, the vapor *implodes* into liquid. If the implosion occurs at or near a metallic surface, damage will occur.

Cavitation can cause rapid failure of the control valve, as well as the piping adjacent to the valve. In this worst condition, cavitation produces sounds often compared to *gravel* flowing through the system. Cavitation is predictable and can be prevented by using correct system design. Refer to a more detailed source for additional information.

Feedback. Proper instrumentation and control systems require information, such as valve position. Feedback from a valve can be presented in discrete and continuous forms. Discrete positions are usually provided by limit switches mounted to trip at the open and close positions. Modulating positioners normally use a potentiometer mechanically coupled to the valve shaft or stem.

Failsafe. System designs should include provisions for loss of power. Power loss can be signal power or primary power. Loss of signal, if the primary power is still online, can be detected by the signal conditioner and a default state can be forced (open, closed, as-is). If primary power is lost, auxiliary power must be provided if operation of the valve under power-fail conditions is required. For electric actuators, battery backup generators are often used. For pneumatic and hydraulic systems, this can be by spring-power or a backup pressurized accumulator.

VALVE SUMMARY

Considerations for valve selection are

- Process fluid to be controlled (chemical properties, temperature, pressure, flow)
- Parameter to be controlled (flow, pressure)
- Control characteristic required (quick opening, linear, or equal percentage)
- System characteristics (line size, adjacent piping)
- Actuating medium and range (electric, pneumatic, or hydraulic)
- Speed of operation (open speed, close speed, multispeed)
- Control signal type and range (3 to 15 psi air, 4 to 20 mA electronic, digital)
- Failure mode (open, close, as-is)

Other considerations are

- Existing equipment
- Experience
- Availability
- Maintenance
- Service life

PUMPING SYSTEMS

Pumps are used to add energy to the process. In most public water systems, pumping facilities are used at raw water intakes of surface water, at wells, in treatment facilities, and for creating head (pressure) in the distribution system. Liquid chemical feeders also use pumps, albeit relatively small ones, and will be discussed in this section. Although conveyor belts can be thought of as pumping solid materials, such as lime or powdered activated carbon, they are discussed later in this chapter under Miscellaneous Final Control Elements.

While in most cases the energy added by pumping is used to push liquid through pipes, sludge collectors in the bottom of sedimentation basins also operate as a special kind of pump. The mechanism pushes the heavier sludge across the bottom of the basin. Similarly, mechanical scum collectors push the lighter scum across the top of sedimentation basins.

Pumps are also used for mixing. They add the mixing energy to various processes in water treatment. Chemicals are often mixed in a batch before injecting them into the flow and after the point of injection to enhance distribution in the process. Flocculators are another example of pumps used for mixing although the mixing is extremely gentle.

A pump installation consists of three major components: the pump, the driver, and the controls. Depending on the application and the type of pump and driver, various types of controls are used. Several methods of controlling pumps with an electric motor were discussed in chapter 3. These may vary from simple on–off switches activated by a timer or a pressure or level sensor through variable speed drives.

For control purposes, the two basic types of pumps used in water systems are displacement pumps and nondisplacement pumps. The amount of energy that displacement pumps deliver to the process is a function of displacement volume and the speed of the pump. The physical properties of the liquid and the system head or pressure have almost no effect on the amount of net energy delivered to the process. It is for this reason that displacement pumps are used for metering pumps in chemical feeders.

Nondisplacement pumps do not have a specific displacement, so the energy they impart is related to the speed of the pump and the physical properties of the liquid and the system head or pressure. Because all pumps can be controlled by varying their speed, this common control method is discussed first. Then control methods specific to each type are discussed.

Speed Control

Speed control of pumps can be applied to the driver itself or to the coupling between the driver and the pump.

Variable speed drives. Variable speed drives are often abbreviated VSDs. These can be either electric motors or mechanical engines. Electric motor VSDs are discussed in chapter 3, Motor Controls.

Because of their high cost and high maintenance, engines are usually used only to backup electrical motors and in remote or portable applications. Engines achieve variable speed through the use of throttles that regulate the amount of fuel fed to the engines. Automatic control uses a positioner similar to a valve positioner (discussed under Valves) to adjust the throttle. These mechanisms operate similar to the cruise control in an automobile.

Mechanical engines can be reciprocating engines or turbines. Reciprocating engines can use natural gas, propane, gasoline, or diesel oil for fuel. Turbines are typically powered by steam or natural gas.

Variable speed couplings. Until the development of the electric VSD, variable speed couplings were used between fixed-speed electric motors and pumps. These couplings used several different technologies. The most common ones are briefly described below.

Variable ratio pulleys are one type of variable speed couplings. They employ a V-notch pulley that can be mechanically widened or squeezed together to change the effective diameter of the pulley. Noncontinuous versions of this type of coupling use several pulleys and function similarly to a 10-speed bicycle.

Eddy current drives are the most common type of continuously adjustable couplings. They use an electromagnet to create a magnetic coupling. These units typically accept a 4–20 mA DC control signal that varies the current in the electromagnet, causing the torque transfer to vary. Therefore, while the input shaft turns at a fixed speed, the output shaft turns at something less than full speed because of slip in the magnetic coupling. Eddy current drives were commonly used on pumps up to 1,000 hp but are now being replaced by the more efficient and reliable electric VSD.

Hydroviscous drives are essentially hydraulic transmissions. They are used on very large pumps because they can handle more torque than eddy current drives. They adjust the torque transfer by moving disks closer or farther apart in a chamber filled with hydraulic oil. As with the eddy current drives, the input shaft rotates at the fixed speed of the motor. The output shaft (and therefore, the pump) rotates at less than full speed depending on the spacing of the disks. These units generate a lot of heat and are not very efficient. Modern electric VSDs are now available for medium- and high-voltage motors, so hydroviscous drives are seldom used in new facilities.

Displacement Pumps

Displacement pumps are used to precisely feed chemicals and to pump highly viscous fluids, such as sludges and slurries. Displacement pumps come in three basic varieties: piston, diaphragm, and rotary. Each of these displacement pumps can be controlled by speed. Piston and diaphragm pumps can also be controlled by varying the volume of the displacement of the pump.

Piston pumps. Displacement of a piston pump depends on the bore (diameter of the piston or cylinder) and the stroke length of travel of the piston. Stroke is more common, but bore is also adjustable on some small models used in chemical feeders. Both bore and stroke are adjusted by varying the length of a mechanical lever. The adjustment is usually done manually via a micrometer knob. The dosage of a chemical is often controlled by manually adjusting the stroke, while the pump speed is continuously adjusted based on plant flow. Some units do have automatically adjusted stroke lengths. These units use a device similar to a valve positioner (discussed under Valves) to accomplish this.

Diaphragm pumps. Diaphragm pumps do not have adjustable bores but do have adjustable strokes. Methods for adjusting the stroke are the same as for piston pumps, although the range of adjustment is relatively small for diaphragm pumps compared to piston pumps.

Rotary displacement pumps. Rotary displacement pumps are typically used to pump highly viscous fluids. These pumps come in several varieties based on the type of rotor used. Common types include gear, lobe, and screw pumps. They differ from the other displacement pumps in that they do not reciprocate. Consequently, they do not have stroke adjustments. The displacement of these pumps is fixed and cannot be adjusted. Therefore, speed control is the only option for controlling these pumps.

Nondisplacement Pumps

Nondisplacement pumps come in three basic varieties: centrifugal, propeller, and turbine. Nondisplacement pumps are the most common and are used for general applications requiring the movement of liquid and high-volume, low-pressure gas. Each of these nondisplacement pumps can be controlled by speed as discussed above. However, because the amount of energy they impart is also a function of the system head, the energy they add can be controlled by adjusting the discharge pressure. This adjustment is accomplished by using a control valve in the pump discharge piping.

The valve can be in series with the discharge pipe and add or remove pressure from the pump's discharge (Figure 8-14). While this configuration may appear to waste energy, at higher discharge pressures (valve nearly closed), the pump will not add much energy due to the back pressure caused by the valve. Therefore, only the small amount of energy transferred to heat is actually wasted. This method is commonly used for flow control.

The valve can also be configured in a bypass line, wasting the energy back to pump suction (Figure 8-15). This method is more wasteful of energy because it allows the pump to add a high amount of energy and then dumps a portion back to the supply or suction side. However, this method is advantageous for pressure control in dynamic systems because of its ability to react quickly and is commonly used in blower controls.

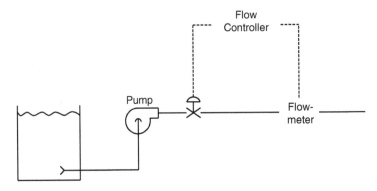

Figure 8-14 Discharge pressure control via series valve

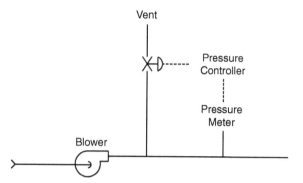

Figure 8-15   Discharge pressure control via bypass valve

MISCELLANEOUS FINAL CONTROL ELEMENTS

Several other FCEs, besides valves and pumps, are used in water treatment plants. These include chemical conveyors and feeders. Selecting the most efficient and effective control equipment that meets treatment requirements will reduce operating costs.

Chemical Conveyors

In large water treatment plants, dry chemicals are moved from the point of delivery to storage facilities and from storage to point of usage by conveyors. Conveyors may be a simple gravity slide controlled by manual gates to a remote-controlled pneumatic system. Depending on the design of the plant, belt or screw conveyors may also be used. Most mechanical conveyors have a single speed and are operated manually or by a remote sensor installed in a storage bin or process tank. Some conveyors are equipped with automatic shutdowns for personnel. Refer to Figure 8-16 for a typical arrangement.

Chemical Feeders

Chemicals used in water treatment can be in either liquid or dry form. Liquid chemical feed pumps have been discussed in this chapter under the section on displacement pumps. A typical total liquid feed system is shown in Figure 8-17. In this diagram, the chemical solution is transferred to the day tank either manually or automatically in a batch mode using the transfer pump. The solution feeder continuously draws the chemical from the day tank and feeds it to the process. The feed rate is adjustable, either manually or automatically, with the solution feeder.

Dry chemical feed systems use granules or nuggets of chemicals and can be designed to utilize either bag storage or bulk storage. A typical dry chemical feed system is shown in Figure 8-18.

Two basic types of dry chemical feeders are gravimetric feeders and volumetric feeders. Gravimetric feeders deliver chemicals based on weight. Volumetric feeders deliver chemicals based on volume. Most volumetric feeders are positive displacement feeders and tend to be less expensive and less accurate than gravimetric feeders. Volumetric feeders are adjusted by varying the speed of the feeding mechanism.

158 INSTRUMENTATION AND CONTROL

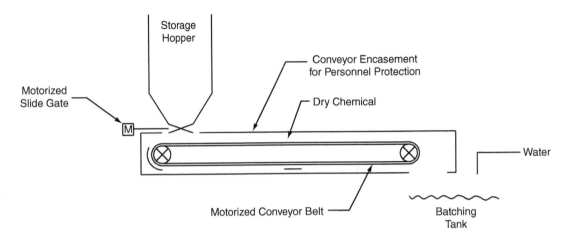

Figure 8-16 Pneumatic conveying system

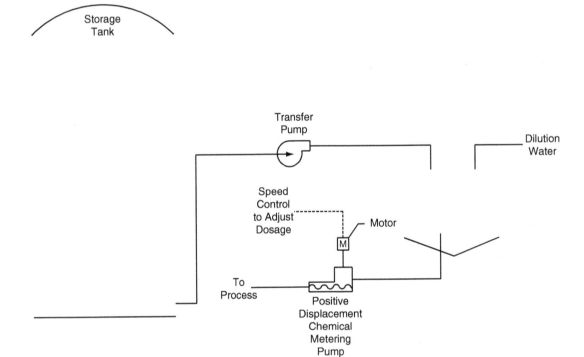

Figure 8-17 Chemical feed system (liquid)

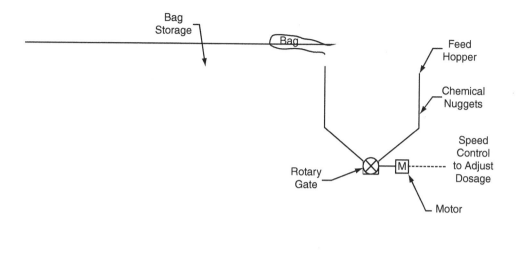

Figure 8-18 Chemical feed system (dry)

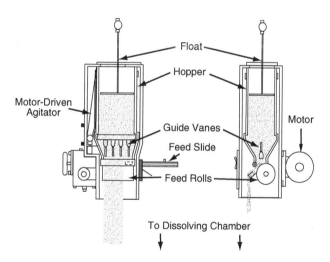

Figure 8-19 Typical rotary paddle volumetric feeder

Two common types of volumetric feeders are the rotary paddle (Figure 8-19) and the screw (Figure 8-20). Other types of volumetric feeders include oscillating hopper feeders and grooved disc feeders. The major drawback to volumetric feeders is an inability to adjust for changes in bulk density of the chemical being fed.

Belt-type gravimetric feeders (Figure 8-21) can usually be used for any dry chemical. The gravimeter belt feeder uses a basic belt feeder with the addition of a control and weighting system. Feed rates are adjusted by modification of the belt speed or the amount of chemical released to the belt. Gravimetric feeders require regular shutdowns for cleaning and calibration to ensure accurate feed rates.

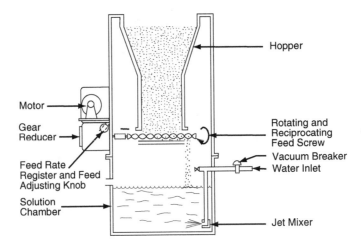

Figure 8-20   Screw-type volumetric feeder

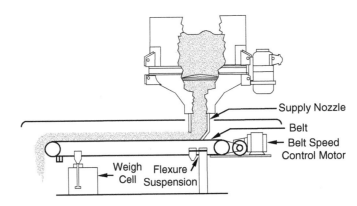

Figure 8-21   Gravimetric feeder (belt type)

AWWA MANUAL

Chapter 9

Basics of Automatic Process Control

Automatic control is a form of process control. Process control is used to improve the quality of water treatment and distribution processes. Important process parameters are measured and certain operating conditions are adjusted to force the process parameter to achieve a desired value. If an operator makes the adjustment, the control is manual. If the adjustment is made with an electrical or mechanical controller, control is automatic.

This chapter discusses common techniques used by controllers to make these adjustments. The intent of this chapter is to provide a basic understanding of process control and discuss the most common techniques used to automate process control. An understanding of how automatic control works can improve operation of a water system.

Automatic controllers attempt to imitate human decision making and action but cannot achieve the level of complexity that humans can. Therefore, automatic control is limited to the more simple process situations. However, many water utility processes are in this category.

As used here, *process* means any single aspect of the water system that needs to be controlled. Common examples are the water level in a reservoir or the pH in a sedimentation tank. When one process follows another, the output of the first process becomes one of the inputs to the next process, as is often the case in water treatment and distribution. Every process has two or more inputs (such as energy or chemicals), a reaction between the inputs, and a desired output.

The collection of equipment used to control each process parameter forms a control loop of information processing. The parameter variables are measured, a decision on appropriate action is made, and then an adjustment to one of the process' inputs is made. The direction of the information flow in the control loop can be either in the same direction as the process called *feedforward*, or in the opposite direction of the process called *feedback*. Figure 9-1 shows these two basic schemes. While either of these can be manually or automatically controlled, they each operate quite

162   INSTRUMENTATION AND CONTROL

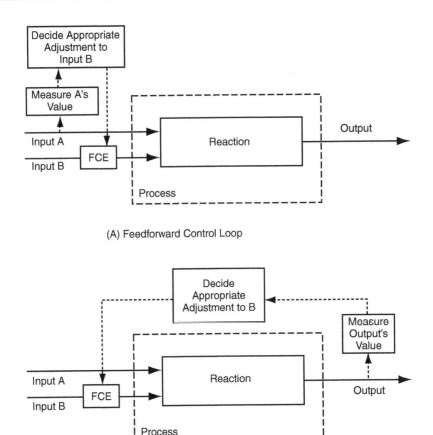

Figure 9-1   Generic control loop

differently. This chapter first examines these two basic types of control, compares them, and then looks at some of the common methods for automating them.

FEEDFORWARD CONTROL

Feedforward control refers to measuring the input values, calculating the proper value of other inputs to obtain the proper output value, and then adjusting those inputs accordingly. To do this successfully, the value of all the inputs that cannot be controlled (called *wild variables*) must be known. How the process works must also be known, and the rest of the inputs have to be adjusted using final control elements.

A feedforward control loop measures one or more inputs of a process, calculates the required value of the other inputs, and then adjusts the other inputs to make the correction. Because feedforward control requires the ability to predict the output, this type of control is sometimes called *predictive control*. Furthermore, because feedforward control does not measure or verify that the result of the adjustment is correct, it is also referred to as *open loop* control. A consequence of open loop control is that if the measurements, calculations, or adjustments are wrong, the control loop cannot correct itself. If the initial adjustment in response to changed conditions does not produce the correct output, the process will continue to produce the wrong output.

BASICS OF AUTOMATIC PROCESS CONTROL

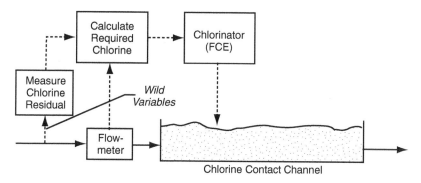

Figure 9-2 Feedforward control of chlorine contact channel

As an example of feedforward control, consider the chlorine contact channel shown in Figure 9-2. Assume a free chlorine concentration of at least 2 ppm in the channel needs to be maintained. If the flow rate and the free chlorine concentration going into the channel are known, the amount of chlorine to add that will raise the concentration can be calculated. If the calculation is performed by an operator who then adjusts the chlorinator manually, feedforward control is used. If an electronic module (or a computer) calculates the required chlorine and adjusts the chlorinator, then control is automatic feedforward. Feedforward control of a process can *only* work if both of the inputs and their reaction are known.

Few processes are so simple that all inputs can be measured and the reaction is well known. Therefore, feedforward control is not viable in most situations. Where feedforward control cannot be used, feedback control must be used.

FEEDBACK CONTROL

Feedback control measures the *output* of a process and adjusts one or more of the inputs to force the output to the desired value. Controlling a process using the feedback method requires three basic steps:

1. The desired process parameter, also called *process variable,* is measured. Examples of process variables are level, pressure, flow, and pH. The particular process variable to be controlled is called the *controlled variable*.

2. The value of the controlled variable is compared to a desired value or *set point*. The difference between the set point and the controlled variable's value is called the *process error*. Process error is calculated by:

   Error = Measured Value − Set Point

   If the controlled variable's measured value is lower than the set point, the error will have a negative value.

3. One of the process inputs is adjusted to cause the controlled variable to change to the value of the set point. The device used to make the adjustment to the input of the process is the final control element.

A feedback control loop measures the output of a process, reacts to an error in the process, and then adjusts an input to make the correction. Therefore, the information loop goes backwards. Because the process only reacts to an error, it is

164 INSTRUMENTATION AND CONTROL

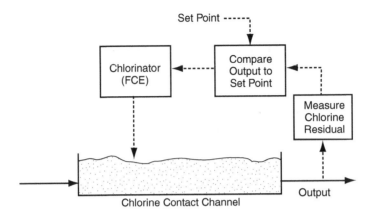

Figure 9-3 Feedback control of chlorine contact channel

also called *reactive control*. Because feedback control does check the results of the adjustment, it is said to be *closed loop* control. Therefore, feedback control, unlike feedforward control, is self-correcting. If the initial adjustment in response to changed conditions does not produce the correct output, the closed loop system can detect the problem and make another adjustment. This process can be repeated as often as necessary until the output is correct.

Feedback control applies to the chlorine contact channel process as shown in Figure 9-3. In this case the free chlorine concentration is measured downstream of the point in the channel where the chlorine is added, and if it is less than 2 ppm, the chlorinator adds chlorine at a faster rate. Conversely, if the chlorine concentration is higher than 2 ppm, the chlorinator can add chlorine at a slower rate.

The difference from feedforward control is that the inputs into the process are unknown. In other words, the output has to be wrong (different than the set point) before the amount of chlorine added will be adjusted. Three or four adjustments may be needed before the set point is reached. Meanwhile, the water going through the system has the wrong chlorine concentration. However, automatic controllers for feedback control do have internal controls, which, when set properly, can minimize the adjustment time. The setting of these internal controls is referred to as *tuning the loop*. Tuning is a very complicated process and should not be attempted by personnel who are not trained in process control loop tuning.

FEEDFORWARD VS. FEEDBACK CONTROL

Under certain process conditions, one of these methods will work better than the other. Feedforward control can inherently provide better continuous control. Feedforward control does not require the process to be wrong before an adjustment is made. However, feedforward control usually requires more instruments than feedback control because feedforward control requires more inputs that must be measured or controlled.

In the chlorine residual process, for example, the feedforward method required two measurements (inlet flow and inlet residual) while the feedback method only required one measurement (outlet residual). Therefore, the advantages of better control available from feedforward control must be weighed against the cost of the extra instruments and control equipment.

Feedforward control also must predict the process reaction to the inputs. The controller must be able to simulate or model the reaction to make feedforward control work. People (in the case of manual control) or a controller (in the case of automatic control) must be able to perform this simulation quickly to make the appropriate adjustments as fast as the process inputs change. For most processes, this is too complicated to do accurately, or the necessary equipment is too expensive. Consequently, even though feedforward control may work better than feedback control, it may be impractical. However, as computers become more powerful and less expensive, the opportunities for successfully implementing feedforward control are increasing.

Feedback control requires the proper adjustment of the controlled input. The problem is then knowing which inputs changed and by how much. The proper adjustment for a given variation in the output is not necessarily the same each time it occurs. As an example, consider the chlorine system again. If the measured channel residual is 1.8 ppm instead of 2 ppm, the negative error is a negative 0.2 ppm. The amount of chlorine that should be added depends on whether the error was caused by a reduction in the chlorine residual of the inlet water *or* an increase in the inlet flow. The reduced inlet residual would only require a small increase in chlorine, while the increase in inlet flow would require a larger increase in chlorine. Because the actual cause is not known, a small adjustment should be made first and then a larger adjustment if the small adjustment did not correct the problem.

In many cases, how the process reacts to some of the inputs is known and easily measured, but measuring all the inputs is not practical. In these cases, feedforward is used for the part of the process that is known and feedback control for the part that is uncertain. This combination of feedforward and feedback is called *compound control*.

For example, consider the chlorine contact channel process example shown in Figure 9-2. If the amount of water going into the channel and chlorine residual are known, the proper amount of chlorine to add could be calculated. But, if the water entering measures zero free chlorine, then the correct amount to add could not be calculated because the chlorine demand is unknown. While chlorine demand can be measured, the equipment is very expensive and complicated to operate. The feedback method shown in Figure 9-3 could be used to avoid this problem. With feedback control, however, if the inlet flow suddenly changes (because another pump was started), improperly chlorinated water will be produced until the low concentration is measured and the correct adjustment is made on the chlorinator. The control of this process could be improved by using feedback control as shown in Figure 9-3. This method would control the chlorinator based on residual after the injection point *and* keeping the feedforward control shown in Figure 9-2 for controlling the chlorinator based on inlet flow. Then the chlorine contact channel's controls would look like Figure 9-4. Compound control often works better than just feedforward or just feedback control but requires more equipment, so the benefit has to outweigh the cost.

MANUAL VS. AUTOMATIC CONTROL

Many measurements are difficult or impossible to make with instruments, and a knowledgeable and experienced person can often do a better job than automatic controllers. If the control is continuous, as is necessary with a fast-changing process, the adjustments must be made quickly and frequently for proper control. To achieve this manually, the operator must constantly watch the measurements and make the adjustments. If the process can be controlled with a relay circuit or an electronic module, then an automatic control system works best.

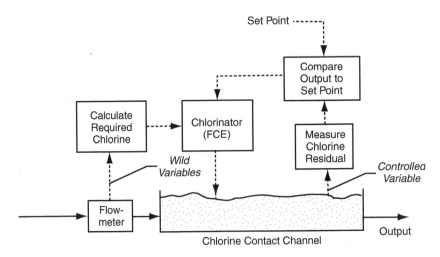

Figure 9-4 Compound control of chlorine contact channel

When computers are used as controllers, complicated mathematics can be performed quickly and automatically. Some computer control programs can actually *learn* to do a better job through experience. These controllers are, of course, expensive and are reserved for the most complicated and most important control tasks. As the price of computers and software comes down, the opportunity to automate more complex processes will increase.

AUTOMATIC FEEDFORWARD CONTROL METHODS

Feedforward control requires the process to be mathematically modeled. Therefore, an automatic controller must calculate what change in controlled inputs will cause the right output. Several devices are commonly available to perform mathematics for use in automatic feedforward control. These devices are available using pneumatic, electronic, or digital-based technologies and can, therefore, be used in any type of control system.

Timers and Event Counters

Timers are devices that keep track of time; event counters count the occurrence of specific process events. The same basic devices are used in both of these cases. These units typically have a switch that allows the machine to count time or number of events. When the count reaches the preset value, the output of the device is triggered.

Timers can be set to trigger their output at a specific time of the day. These are called *time-of-day* timers. Alternatively, they can measure elapsed time and trigger their output after a preset amount of time has elapsed after some event. These are called *elapsed-time* timers. Any process or equipment switch can be used to start the time measurement. For example, rake drives are run in sedimentation tanks for a short period of time every day, and blowdown valves are open for a short period of time after the rakes have finished. A 24-hour time-of-day timer could then be set to turn the rake drive motor on at 2:30 p.m. every day and off at 3:30 p.m. every day. A rake drive shutting off could trigger an elapsed-time timer to count twenty minutes and then open the blowdown valve. Another elapsed-time timer could be triggered to start timing when the blowdown valve is opened, then count five minutes of elapsed time before shutting the blowdown valve. Numerous other possibilities exist because timing is a major method of controlling processes in a water utility.

Event counters are referred to as *up counters* or *down counters*. Up counters always start at zero and count up to the desired number, while down counters start with the desired number and count down to zero. One very common example is a controlled sampler. Samplers are often required to take a sample after every 1,000 gallons of water. A flowmeter can put out a pulse (by closing an electrical contract briefly) after every 100 gallons. The counter could be set to 10 when the sampler is triggered. As with timers, numerous possibilities exist.

Function Modules

Function modules are a general class of instrument that can perform simple arithmetic and logic functions. Arithmetic functions include everything on a common handheld calculator, such as addition, subtraction, multiplication, division, or square root. Logic functions include the common Boolean algebra operands such as AND, OR, or NOT. These devices can be used in various combinations to execute the required calculations for feedforward control. Function modules usually accept one to four input signals and provide one output signal.

If the required equation includes a parameter that the operator needs to adjust from time to time, then a knob or switch has to be provided and connected to one of the module's inputs. The knob or switch must also have a scale or nameplate so the operator can determine what value is being selected.

Ratio and Bias Controllers

Ratio and bias control are very common types of feedforward control. They both execute the following general equation:

$$Y = mX + b \qquad (9\text{-}1)$$

Where:

Y is the output
m is a constant whose value is multiplied times the value of input X
X is the wild process input variable
b is a constant whose value is added to the product of m and X

When the value of m is operator-adjusted and the value of b is fixed, the control is called *ratio control*. Conversely, if the value of m is fixed and the value of b is operator-adjusted, then the control is called *bias control*.

This equation results in a straight line. Consequently, any process in which the output is linearly related to one of the inputs can use this equation to model the process and predict the output.

A linear relationship is common in process control, and controllers are manufactured to expressly execute this equation by incorporating the necessary function modules, scales, and knobs all in one dedicated instrument. Executing the ratio or bias form of the equation can be chosen with a switch on the instrument. When ratio is selected, the front panel knob and scale determine the value of m, and an internal adjustment determines the value of b. Conversely, when bias is selected, the front panel knob and scale determine the value of b, and an internal adjustment determines the value of m. Use of the bias equation is much more common in the food processing and manufacturing industries than in water utilities. Ratio control is the most common method for controlling chemical addition in water treatment facilities.

For example, the amount of chlorine needed to maintain any particular concentration is linearly related to the amount of flow. Therefore, if the flow doubles, the chlorine injection rate doubles. A ratio controller can accomplish this by using the flowmeter's output signal as X and the front panel knob's position as the value of m and the internal value of b set to zero. The output signal Y is then used to control the chlorinator. The front panel scale is calibrated in parts per million so the operator can easily dial in the desired concentration or dosage. Ratio control used in this way is often referred to as *pacing*. The chlorine is said to be *paced* on plant flow. With automatic ratio control, every time the plant flow changes, the chlorinator's feed rate will also change appropriately.

Computers

Computers can also perform the calculations required for feedforward control. Because computers cost more than timers, function modules, or ratio–bias controllers, they are rarely dedicated to a single application. Efficient use of computers usually requires that they perform the required calculations for several control loops at the same time. However, smaller and less expensive computers are now available that can be put into the old ratio–bias controller's case and perform the same function economically.

Computers can handle several loops at the same time with only a marginal increase in the cost. They can also perform very complicated calculations and have become very popular for use as feedforward controllers. Where several control loops are needed, the efficiency of the computer makes it the best choice for implementing the function of the feedforward controller on all the feedforward control loops.

AUTOMATIC FEEDBACK CONTROL METHODS

The relationship between the actuation of the final control element (e.g., the chlorinator) and the changes in the controlled variable (e.g., free chlorine concentration) is referred to as the *control mode* or the type of control action being used.

Four important types of control action are regarded as basic control modes: on–off, proportional, integral (or reset), and derivative (or rate). These basic control actions can be used individually as well as being combined together to form the kind of control action needed.

In describing these control modes, graphs show how the output of the controller changes over time as the value of the controlled variable moves above and below the set point (Figure 9-5).

In this manual, the value of the controller's output will always be plotted on the upper graph's Y, or vertical, axis. The value of the controlled variable and the set point is plotted on the lower graph's vertical axis. Time will always be plotted on the horizontal axis common to both of these graphs. Specific points in time are referenced by a dashed vertical line labeled with a lowercase letter on the time axis.

On–Off Control

The simplest and probably oldest form of feedback control is on–off control. This mode of control changes the value of the controller output from one position to the other as the measured value of the controlled variable goes above and below the set point. This control action is, consequently, limited to making very coarse adjustments.

Figure 9-6a illustrates the on–off control of water level in a water distribution system reservoir, and Figure 9-6b shows the timing graph for the controller's operation. The pump is turned on or off depending on the level in the tank compared

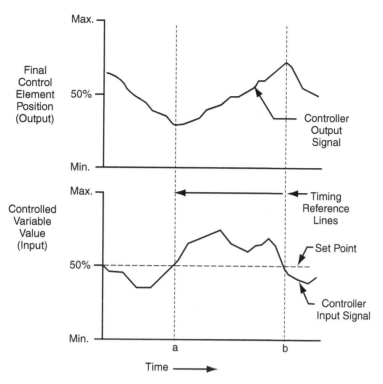

Figure 9-5 Generic feedback control timing graph

to the level set point. Here the set point is actually a range of acceptable values rather than a precise point. The range is determined by the position of the tank's high-level switch and a low-level switch. As the flow into the tank causes the level to rise above the high-level switch (point [a] in Figure 9-6b), the pump is turned off and only goes on again after the level falls below the lower level switch (point [b] in Figure 9-6b). A circuit for achieving this very common and simple type of control is provided in chapter 3, Motor Controls.

While on–off control can work very well, it lacks the ability to easily change the range of level that is maintained in the tank. To have operator control on the range of level, a controller known as a gap-action or on–off differential controller is used. Instead of using level switches to start and stop the pump, a controller compares an analog measurement of the level to an operator set point. The controller then switches the state of its two output contacts to effect motor starting and stopping or valve opening and closing. Alternatively, the controller can generate a continuous analog output signal, such as 4–20 mA DC, and vary the control output from one extreme to the other, that is, from 0 to 100 percent.

An interval, known as a *differential gap*, is established above and below the set point where the final control element does not change. This gap is analogous to the distance between the two level switches in Figure 9-6a. The control action only takes place when the controlled variable is above or below the specified gap around the set point. As long as the measurement remains within the gap, the controller holds the last output state. A very common example of an on–off differential controller is a heater thermostat. If the measured temperature is below the set point, the thermostat turns the heater on. When the temperature rises to the set point, the thermostat turns the heater off. To prevent excessive cycling of the heater on and off

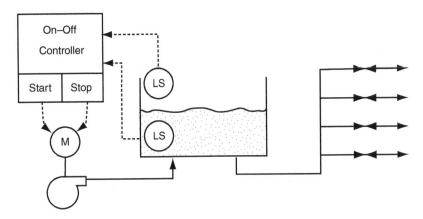

Figure 9-6a   On–off control of a reservoir

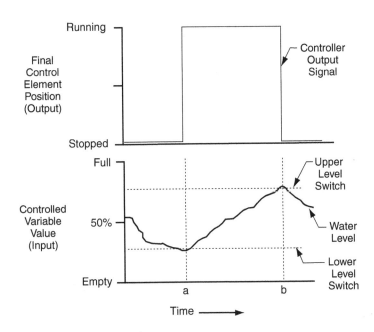

Figure 9-6b   On–off control timing graph

around the set point, there is usually a small dead zone or gap of a couple of degrees around the set point that the thermostat will not switch. The dead zone causes the heater to go on at a temperature slightly below the set point and off at a temperature slightly above the set point.

Figure 9-7a shows the application of gap-action control in a reservoir level control loop, and Figure 9-7b shows the timing graph for this type of control. With this type of control, the level in the tank is measured with an analog level transmitter instead of the level switches, as with on–off control. The controller will maintain the level between the top and the bottom of the gap. The width of the gap is adjustable and the center of the gap is the set point entered on the front of the controller. As the level recedes to the bottom of the gap (point [a] in Figure 9-7b) the

BASICS OF AUTOMATIC PROCESS CONTROL   171

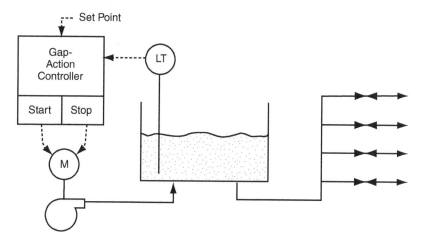

Figure 9-7a   Gap-action control of a reservoir

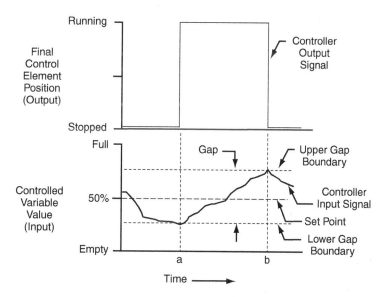

Figure 9-7b   Gap-action control timing graph

controller turns the pump motor on. This causes the level to rise until the top of the gap is reached (point [b] in Figure 9-7b) when the controller will turn the pump off.

Proportional Control

Proportional controllers adjust their output signal in proportion to the value of their input signal. A one-to-one relationship always exists between the final control element position and the controlled variables' value. Proportional controllers always have an internal gain adjustment that determines the change in the output signal resulting from a given change in the input signal. This adjustment is called the *proportional band* (PB). PB is the percentage of input variable change that will cause the output signal to vary from 0 to 100 percent. Some controllers have a *gain*

172   INSTRUMENTATION AND CONTROL

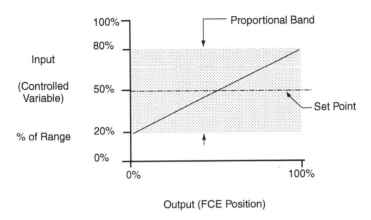

Figure 9-8   Proportional control input/output relationship

adjustment instead of a proportional band adjustment, which is the same thing expressed differently. Gain is the inverse of proportional band.

The center of the proportional band is the set point. Half the band exists above the set point and half of it exists below the set point. The proportional band functions similarly to the gap in the gap-action controller. However, unlike the gap controller, a proportional controller must use a final control element that is a continuously adjustable device, like a motor speed controller or a valve positioner.

For example, if a proportional controller has a set point of 50 percent and a proportional band of 60, then the band will exist from 20 percent to 80 percent of the controlled variable's range. This concept is illustrated by the graph of the controller's input versus its output, as shown in Figure 9-8.

If the input signal (the controlled variable's value) is at the set point, then the output signal (i.e., the final control element) will be at its midpoint. If the controlled variable's value drops to 20 percent of its range, the output signal (and therefore, the position of the final control element) will change to 0 percent (its minimum position). Alternatively, if the controlled variable's value rises to 80 percent of its range, then the output signal (and therefore, the position of the final control element) will change to 100 percent (its maximum position). The controlled variable will always be maintained within the PB. The PB can usually be adjusted to any value between 5 and 200. Smaller PBs will result in tighter control (i.e., narrower bands) but will also cause more movement of the final control element, which will wear it out faster.

Proportional control is the feedback control version of the feedforward control method called *ratio* control. The output is the result of multiplying the controlled variable's value by a constant determined by the PB and adding a bias determined by the setting of the set point.

In the reservoir level control example, Figure 9-9a shows the application of proportional control to reservoir level, and Figure 9-9b shows the timing graphs for this type of control. Assume the level is at the set point, and the current motor speed is pumping as much water into the reservoir as the customers are using. If the customer demand suddenly goes up, the level in the reservoir will begin to recede (point [a] in Figure 9-9b). As the level recedes, the output of the proportional controller begins going up. The pump speeds up so water is pumped in at a higher rate. Eventually the pump will reach a speed that causes the inlet flow rate to match the outlet flow rate. The level will stop receding and hold steady (point [b] in Figure 9-9b). Because the level is not changing, the controller will not change the

BASICS OF AUTOMATIC PROCESS CONTROL 173

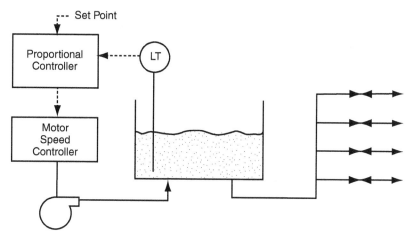

Figure 9-9a Proportional control of a reservoir

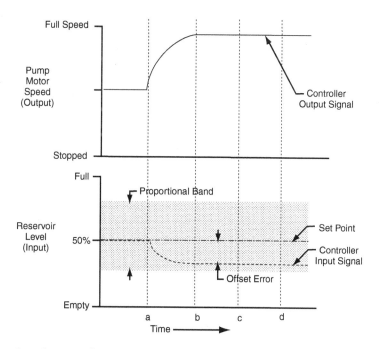

Figure 9-9b Proportional control timing graph

output further. Consequently, the system is now in a stable condition. However, the level is not at the set point because of an offset error. The occurrence of this offset error is an inherent characteristic of proportional control.

The offset error can be overcome by adjusting the set point to a value above or below the desired value. Suppose the process has a value of 30 percent when the set point is 35 percent. The offset error is causing the process to be 5 percent below the set point. If the set point is changed to 40 percent, the offset error will cause the process to be at 35 percent, the original set point. This adjustment of the set point is referred to as *resetting the set point*, and because the operator has to do it, it is called *manual reset*. If the reset only had to be done once, it would not cause problems, but the amount of offset error varies with each change in the process' inputs. Therefore, to eliminate offset error using manual reset, the set point must be reset every time the process inputs change.

Several things should be noted about the proportional control system:

- The controlled variable is tied to the motor speed such that for every position of the level (or controlled variable), there is a corresponding value of the motor speed (or final control element).

- The ratio between the motor speed and the tank level that can be adjusted is called proportional band or gain adjustment.

- There will always be an offset error, and this is usually acceptable on applications such as level control. However, when a proportional controller is applied on flow or temperature, an offset error may not be acceptable.

- Offset error can be overcome by manually resetting the set point to a slightly different value so that the actual value (i.e., set point plus offset error) will be the desired value.

Automatic Reset or Integral Control

To reduce the need for manually resetting the set point, automatic reset control was developed. Automatic reset control is also known as *integral control* because integral calculus is used. Integral control is based on the principle that the controller's output should be proportional to both the size and duration of the error. The controller's output will continue to change its value until the error is zero. This property enables integral action to eliminate offset error. Integral action will only be stable when the measurement has returned to the set point. As long as an error exists, integral action will drive the output in the direction that reduces error.

To regulate the speed at which integral action will drive the output in the direction that reduces error, integral controllers have an adjustment known as *reset* or *integral rate*. This adjustment is measured in *repeats-per-minute*. The term refers to the number of times the controller will cause the output to change by the same amount as the proportional band or gain adjustment per minute of time. The bigger the number, the faster the controller will move the final control element. Because integral control action adjusts speed of the adjustment, it is sometimes referred to as *proportional speed floating control*. The term *floating*, as used here, refers to the fact that the actual value of the controller's output is not predictable as with proportional control. That is, no specific predictable relationship exists between the controlled variable's value and the position of the final control element.

Figure 9-10a illustrates the concept of integral control applied to the reservoir example. Again, assume the level is at the set point, and the current motor speed is pumping as much water into the reservoir as the customers are using. If the customer demand suddenly goes up, the level in the reservoir will begin to recede (point [a] in Figure 9-10b). As the level recedes, an error causes the output of the integral controller to begin rising. This action increases the pump speed so water is pumped in at a higher rate. Eventually, the pump will reach a speed that causes the inlet flow rate to match the outlet flow rate. Then the level will stop receding and hold steady as with proportional control (point [b] in Figure 9-10b). However, because the integral controller is reacting to error and not a process value, it will continue to speed up the pump. The inlet flow rate will now exceed the outlet flow rate, which will cause the level to rise. As the level rises, the error will reduce until the level reaches the set point (point [c] in Figure 9-10b). The tank level is now at the set

BASICS OF AUTOMATIC PROCESS CONTROL 175

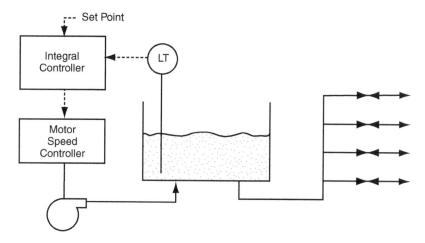

Figure 9-10a   Integral control of a reservoir

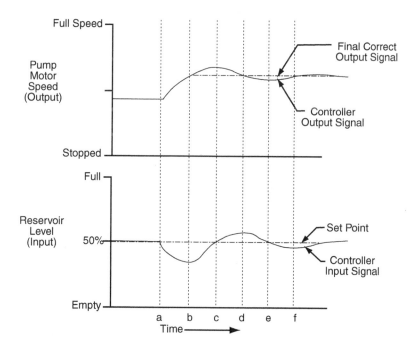

Figure 9-10b   Integral control timing graph

point, but the pump is still pumping more water in than the customers are using. Therefore, the level will continue to rise. The error will now build up in the opposite direction, causing the controller to reduce the pump's speed. Eventually, the pump speed will be reduced so that the inlet flow rate once more equals the outlet flow rate (point [d] in Figure 9-10b). Now the pump is at the correct speed, but the level is slightly high, so there is still an error, and the controller will continue to slow down the pump. This will cause the level to recede below the set point again and begin the whole cycle over. However, if the integral rate is chosen properly, the next cycle will be smaller and eventually, the variations will be too small to notice.

This cycling of the process in response to integral action must decay (reduction of amplitude) for proper control. If the cycling does not decay but rather causes bigger swings in each cycle, the process will be out of control or unstable. An unstable process can be disastrous. In this example, the reservoir will eventually be alternately drained dry and overflow on each successive cycle. Clearly, this condition must be avoided. Faster integral rates will adjust to a process change more rapidly than slower rates, but faster rates run the risk of losing control of the process under certain conditions. Therefore, integral control is usually adjusted to provide a relatively slow recovery from process error. To avoid possible injury to personnel and damage to equipment, the adjustment of the integral rate must only be made by a properly trained control system technician.

No corresponding value of the motor speed for each value of the controlled variable is available as with proportional control. The motor speed will float to the value required to bring the controlled variable back to the set point.

Proportional-plus-Integral Control

Although proportional and integral control action are useful individually, they are most often combined together in what is referred to as *proportional-plus-integral* control, abbreviated as *P+I control,* or simply *PI control*. PI control is one of the most widely used control modes.

The proportional mode has a fast response to most changes and can be used with moderate time lags. The integral mode, while sluggish by itself, compensates for offset error. This type of control *can* be tuned to provide much faster corrections than integral action alone without risking losing control. So, except for cases where there are very rapid disturbances to the process or where there are very large lags (inertia), this type of control mode performs well.

Proportional-plus-Derivative (Rate) Control

The control actions previously discussed depend on the actual measured value of the controlled variable. However, with processes that have a lot of inertia (i.e., tank levels or boiler temperature), the control action will be very sluggish. Both of these methods are insensitive to how fast the process is changing. With large inertia processes, a fast change in load will take a long time to correct if the corrections are small, as will be the case with proportional and integral control. To compensate, rate control was developed. Rate control is more commonly referred to as *derivative control* because differential calculus determines the output. The potential correction is solely determined by the rate at which the error occurs. Derivative control is unaffected by the size or magnitude of the error. If the error is constant, then there is no output from derivative control.

Because the derivative response is unrelated to the absolute value of the measurement, whenever the measurement stops changing, the derivative contribution returns to zero. When it starts to change, derivative action opposes that change whether the measurement is moving away from or toward the set point. Because of this characteristic of giving a corrective signal only when the error is changing, the derivative mode is not satisfactory as a control function by itself.

To better understand why derivative control cannot be used, consider the reservoir example again. This time, assume that the system is stable at the set point and nobody is using water in the service area. No water is going out and no water is being pumped into the reservoir. Since the level isn't changing, there will be no output from the controller and, therefore, the pump will be at zero speed. When someone starts using a large amount of water—a fire hose, for example—the outlet

BASICS OF AUTOMATIC PROCESS CONTROL 177

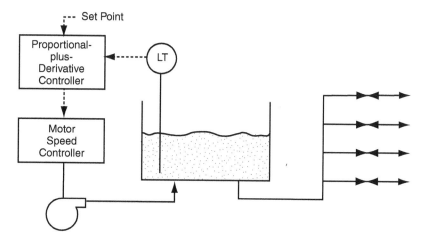

Figure 9-11a Proportional-plus-derivative control of a reservoir

flow will increase, and the level will recede rapidly. Derivative control will increase the pump's speed in direct proportion to the speed at which the level is falling. Consequently, the fast decline in reservoir level will cause a fast increase in pump speed. As more water is pumped into the reservoir, the level will stop falling quickly. The controller's output will then become small because the level will now change slowly. However, a low controller output causes the pump to slow down, which results in, less water going in and the level will begin to recede faster again. The rapid decline in level causes the controller output to go to a high value again. High pump speed causes the water level to stop falling as quickly, but this results in the controller reducing its output and slowing down the pumps again. This control loop is out of control and will cycle the pump speed from zero to maximum continuously. Clearly, derivative control by itself is unsuitable.

While derivative control action is not useful alone, its quick response and ability to oppose changes in the controlled variable can enhance the operation of proportional control. The quick response can reduce the offset error inherent in proportional control by allowing the process to adjust more quickly, thereby reducing the time for the system to drift off the set point. Controllers that provide these two types of control action are referred to as *proportional-plus-derivative* controllers, abbreviated *P+D*, or simply *PD controllers* (Figure 9-11a).

To better understand how derivative control can enhance proportional control, consider the reservoir example again. As before, assume that the system is stable at the set point and a fire breaks out in the service area. Firefighters connect their hoses and turn them on at point (a) in Figure 9-11b. A large increase to the outlet flow will cause the level to recede rapidly. While proportional control action alone will eventually increase the pump speed to match the demand, the offset error will be quite large. The reservoir level will be quite low when the pump finally catches up to the demand, as shown in Figure 9-11b by the dashed horizontal line in the timing graph.

When derivative control is also used, the derivative control action will also increase the pump's speed in direct proportion to the speed at which the level falls. Consequently, the sum of both control actions after point (a) will cause a fast increase in pump speed. The amount of the change in pump speed will depend on how fast the level recedes and on the derivative time constant entered into the controller. The effect of both control actions, proportional plus derivative, causes the sudden increase in pump speed that provides more than the correct increase to match inlet flow. The

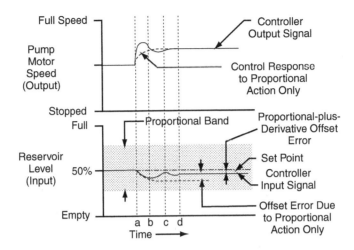

Figure 9-11b Proportional-plus-derivative control timing graph

decline in level will quickly slow down, stopping at point (b) in Figure 9-11b, and actually start to rise. However, even though the derivative adjustment becomes very small as the level stabilizes, the proportional control action is still active and will continue to slow the pump speed gradually. As the level rises between points (b) and (c), the derivative action begins again and helps the proportional action reduce pump speed. This cycle will repeat until the pumped inlet flow eventually matches the outlet flow as at point (d). The level in the reservoir will still have an offset error, but it will be smaller using proportional control alone. The reduced offset results from initially increasing pump speed faster and therefore, stopping the level from receding more rapidly. Without the derivative control boost at the beginning, the proportional control alone would not have caused increased pump speed and matched outlet flow until point (c) in Figure 9-11b. Therefore, the benefit of adding derivative control to the reservoir was a faster reaction to the large demand change and a smaller offset once the system reached stable operation.

Proportional-plus-Integral-plus-Derivative Control

Proportional-plus-integral-plus-derivative control mode provides the best control possible using conventional equipment. The advantages of each are retained and direct, proportional correction of the proportional mode is supplemented by the offset eliminating or reset nature of the integral mode and the stabilizing, quick-acting nature of the derivative mode which is effective in overcoming all forms of lags. Modern controllers typically provide all three control actions in a single instrument. These are referred to as *PID* controllers. Computers also typically provide all three control modes for use in feedback control loops.

REFERENCES

Bogart, T. F. Jr., P. E. 1982. *Laplace Transforms and Control Systems Theory for Technology*. New York City, NY: John Wiley and Sons. Pages 163–210.

Considine, D. M. 1974. *Process Instruments and Controls Handbook*. New York City, NY: McGraw-Hill, Inc. Chapters 17 and 18.

Johnson, C. D. 1982. *Process Control Instrumentation Technology*. New York City, NY: John Wiley and Sons. Pages 279–349.

Tucker, G. K., and D. M. Wills. 1962. *A Simplified Technique of Control System Engineering. Graphical Methods of Understanding and Improving Process Control*. Honeywell, Inc. Pages 1–28.

AWWA MANUAL M2

Chapter 10

Digital Control and Communication Systems

The computational speed and memory capacity of digital computers enable system operators to improve process control. Examples of how computer technology can improve process control include:

- Computerized control allows automation of routine operator tasks, freeing operators to concentrate on exceptions.

- Computerized automation permits more precise control of process variables.

- Much of the process control can be centralized, which allows operators to obtain process information efficiently and implement decisions effectively.

- Many emergency situations can be resolved by prespecified corrective actions.

Digital technology continues to improve, increasing the effectiveness of computer control. Some examples are

- Cost of computer hardware continues to decrease despite impressive increases in capability and speed.

- User implementation is made easier by the development of high-level languages that incorporate built-in procedures, minimizing end-user programming requirements. Graphical user interfaces, which use icons and menus to make operations more intuitive, allow users to concentrate on the process rather than the computer.

This chapter presents overall concepts and guidelines to aid in the understanding of digital control in water system applications. As with many topics in this manual, digital control is too complex to be covered completely. The references listed at the end of the chapter provide additional information.

DIGITAL CONTROL SYSTEMS

A digital control system consists of hardware and software. Hardware is the computer and peripheral equipment, such as electronic and electromechanical equipment designed to treat electrical impulses coded as numbers. Software is a set of coded instructions operating on those numbers and controlling the hardware.

Two unique features set digital control systems apart from other data processing computer systems: 1) real-time operation, and 2) handling of process input/output (I/O). Because water treatment and distribution are continuous processes, constant monitoring and supervision is required, and the computer must operate continuously. It must maintain a constant watch on the process in real time and respond with appropriate control outputs. Therefore, digital process control systems require computers that run real-time operating system software.

Real-time software divides control commands into modules that can be sequentially executed for various functions. Real-time software provides communication between modules. The job of a computer in a large process control system could be likened to a skilled juggler with ten balls in the air. If the juggler concentrates on one or a few of them for too long, the others will drop. Similarly, the computer must scan inputs and control outputs of up to many thousand points and not lose contact with any for more than a few seconds at a time.

As applied to water systems, real-time processing is defined as the interconnecting of a process (i.e., water treatment or distribution control) with a computer capable of analog-digital and digital-analog interfaces and generalized digital (binary) data interfaces. To acquire data correctly, the computer must be keyed to the time scale of the process. Its response in the form of control outputs must be made as quickly as possible to achieve the desired effect on the water system. Because the digital control systems are often unattended, the software used must be reliable.

A single-processor digital computer is, fundamentally, a serial device and can perform only one operation at a time. By carrying out many thousands of operations per second, one after the other, the single-processor computer can appear to be doing things in parallel.

To perform in real time, all digital computer systems must handle *interrupts*. Interrupts are signals to the processor to indicate a task of high importance needs to be done. Most interrupts are external interrupts, in that they originate outside the processor in peripheral devices such as sensors. I/O operations, the source of most external interrupts in most digital control systems, must be able to *break into* the computer's sequence of operations. The computer then finds out which device caused the interrupt, responds in a timely fashion, and resumes its original operations. This process is called *servicing* the interrupt. Because the digital control systems are often unattended, the software used must be reliable.

One of the principal benefits of computer-based digital control systems is that they can be programmed for specific tasks, and programs can be easily modified either during the development phase or after implementation. A high level of complexity is possible, allowing for sophisticated and accurate control.

A modern plant control system applies real-time digital control system technology to monitor and control water treatment plant processes. Plant parameters typically monitored include: process variables such as flow, rate, loss of head, water quality (i.e., turbidity, pH, or chlorine residual), chemical feed rates, and digital signals (for example, pump on or off and alarm signals). Control outputs start and stop pumps, adjust flow rates, adjust chemical feed rates, and wash filters as needed.

The decision-making logic and computations for control are integrated with those of monitoring, thereby delegating a significant task load to the computer. As its monitoring tasks, the computer continues to scan sensors, evaluate the input data for alarms and status changes, generate displays, and store historical data. For control purposes, the computer's task cycle also includes the evaluation of system status based on input data, determination of whether any control outputs are necessary to initiate a process or to regulate process variables, and to issue corresponding outputs.

Digital technology has allowed design of sophisticated supervisory control and data acquisition (SCADA) systems. A SCADA system is similar in concept and operation to a plant control system, with the added component of long-distance telecommunications. SCADA systems are commonly used in the water industry to control the process of water distribution, which may include extensive water collection and transmission components. They are capable of controlling water treatment processes as well.

The analog inputs to a SCADA system include pressure, level, flow, temperature, rotating equipment speeds, and valve settings. Digital inputs may include alarms such as intrusion, high temperature, or pressure. Control outputs include pump, valve, and pressure-reducing valve control.

A basic digital control system has the following components and tasks (Figure 10-1):

- Sensors provide the conditioned input signals from the process to the control system.

- An input interface converts the conditioned signals into computer-compatible codes as input data.

- The computer processes the input data according to instructions contained in its program and generates output data.

- An output interface converts output data into suitable control signals. These signals may be further processed by output conditioning circuits.

- Actuators produce motion and can be used to drive final control elements acting on the process.

- The keyboard and display allow the operator to communicate with the system and control the process.

The following paragraphs discuss each of the digital control elements in greater depth.

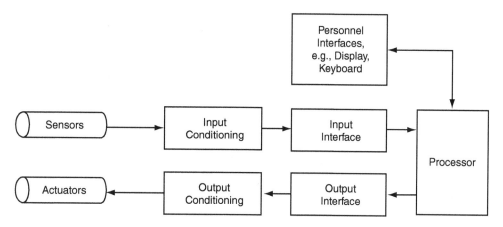

Figure 10-1 Digital control system

Computers

Digital computers are based on microprocessor technology. A microprocessor is a *chip* of silicon using integrated circuit technology developed in the late 1960s and early 1970s. Process control computers include the same basic components as most digital computers, with a few unique custom features designed to enable control of processes. A discussion of the major components follows.

Central processing unit (CPU). The CPU is the executive center of the computer, the hardware associated with the execution of instructions. The CPU has a communication path called a *bus* for transmitting and receiving data. The central processor calls an instruction from memory, interprets and executes it, and then calls the next instruction in the continuous sequence. The central processor controls the activity of the bus, performs computations, and stores the results in the appropriate memory locations.

Main memory. During program execution, a computer's main memory is continually accessed by the CPU for instructions and data to be processed. Each individual location of memory has a unique address, and the contents of each memory location may be either an instruction or an element of information. This arrangement allows the CPU to seek data by address and act on its contents. Main memory, which may be read from and written to in any order, is called random access memory (RAM). RAM consists of integrated circuits and may vary in capacity from a few hundred bytes (one byte is eight bits, or binary digits, of information) to many megabytes (a megabyte is one million bytes). RAM can store variable data or the programs that may be run on the computer. Access time (the time required to retrieve the data for processing by the CPU) to data in RAM is very short.

Many modern computers have volatile RAM; that is, information stored in them is lost when the computer is turned off. Some have battery backup to avoid losing data. Many computers also have a memory section reserved for storing permanent data and programs. This memory is called read-only memory, or ROM. This memory may not be altered during normal operations, nor is it normally volatile.

Mass memory. Mass memory increases the useful data storage capacity beyond that available in main memory. Mass memory is also used to retain backup copies of programs and data when the computer is turned off. Mass memory is usually recorded on a magnetic disk of ferrous-oxide coated materials, much like audio information is stored by a conventional tape recorder. This storage requires some mechanical manipulation to physically locate and access data stored by address, so a relatively longer time for communication to mass memory is required as compared to the purely electronic access of main memory. Communications between the CPU and mass memory may, in fact, be slower by a factor of a thousand or more. Mass memory, while slower than main memory, has the dual advantages of lower cost per unit stored and is not volatile.

Mass memory is commonly used to store database information, archive historical data and reports, store infrequently used programs and programs that are too large to fit entirely into available main memory. When this occurs, the program is segmented into sections, each of which can fit into main memory. The segments are transferred, one at a time as needed, into main memory and executed. Transfer of data and programs into and from mass memory is in blocks. Blocks are large groups of data, which allow sequential instructions by the CPU to be performed as uniformly and efficiently as possible with only a few transfers to and from mass memory.

Input/output system. I/O equipment permits communication with peripheral equipment, which is equipment other than the computer, such as field instruments and control valves. Peripheral equipment is necessary for the computer to do useful work, and the equipment communicates with the computer through connectors called ports. The inputs (measurements and other data) and outputs (control instructions from the computer) are collectively called the I/O portion of the machine.

Computer Peripherals

Computer peripherals include all those components attached to a computer which are used to communicate with the process (water treatment or distribution), people, or off-line storage devices. Communicating with the process is done by I/O subsystems, which are discussed in more detail in the next subsection. Communicating with people is done with a collection of equipment at a common location that is collectively referred to as an operator interface station (OIS).

Some of the major advantages of using a digital control system for plant process control and SCADA are the data processing capability and the variety of data presentations available for operators. The data processing capability provides the automatic generation of all periodic summary reports and can save considerable time and labor previously used for computations, data gathering, organizing, and typing. Data presentations from a digital system permit the elimination of large-scale panels of indicators and recorders. Instead of having to concentrate on a large panel of analog indicators, the operator can access all or most system data with a keyboard and video display unit (VDU).

One of the most effective tools is the graphic display capability of a VDU. A schematic line diagram (commonly referred to as a *graphic*) of a process, such as a treatment plant or pumping station, is programmed to appear at operator command on the VDU. Each signal and its respective value or status is identified and superimposed on the line diagram. As analog values and the status of equipment change in the process, the changes correspondingly appear on the graphic display, providing the operator with an online diagrammed view of the system's current status. The latest digital control systems permit hundreds of user-definable graphic displays to be stored for on-call selection by the operator. Multiple VDUs allow more than one process area to be monitored simultaneously.

Other hardware components (peripheral devices) are also available to communicate with the operator. For example, a screen printer makes a printed copy of whatever is displayed on the VDU. Other developments, such as intelligent terminals (terminals with integral memory and microprocessors); mouse, trackball, and touch screens; and windowing (the ability to display multiple views on a single VDU) have combined to make operators' consoles extremely effective tools for monitoring and supervising water treatment and distribution processes.

In general, the operator interface must be as useful, powerful, timely, and intuitive as possible. Software should enable the operator to navigate from screen to screen throughout the system as quickly and easily as possible and facilitate appropriate response to system demands or urgencies.

Other peripheral devices commonly used as part of a configured system include:

- Additional VDU terminals for programmers to use in entering and editing programs. In some current technology systems, the operator's console terminals also function as programmer's terminals.

- A mass storage system, generally a high-speed, nonremovable disk to hold software files, that exceeds the capacity of the solid-state memory (main memory) of the system

- A lower-speed, removable disk system (e.g., floppy disk, zip, CD drives) which permits backing up programs and files for security and archiving

- Printer(s) to allow production of reports, program listings, and other printed documents

- Optical disk or high-speed cartridge tape drives to facilitate archiving data and reports on permanent media

Process I/O Concepts

The major data forms available to be read (input) or generated (output) by a process control computer are analog, digital, and pulse. These may be monitored points (inputs) from the process to the computer, or they may be control signals (outputs) from the computer to the final control element acting on the process.

Analog data. Analog inputs are measurements of continuous variables, such as flow, pressure, level, pH, turbidity, chlorine residual, and temperature. Through selective sensors and instrumentation (discussed elsewhere in this manual), the variables are detected and converted to electric current signals, commonly in the range of 4–20 mA. These signals are calibrated to an appropriate range of engineering units, such as millions of gallons per day, feet, pounds per square inch, etc. Analog outputs are control signals defined by the computer. These signals may drive a mechanism, such as a strip-chart recorder, or control the process using direct control operation of pump speed or valve position or as a set-point reference for a controller. As continuous variables, analog values are not intelligible to digital computers. Analog values are changed by an analog-to-digital (A/D) converter to a digital (an instantaneous numerical value, usually binary) form that can be stored in the computer. Conversely, a digital-to-analog (D/A) converter converts digital outputs into analog outputs, compatible with the external control device.

Digital data. Digital data exist in discrete, or binary, form (on or off, set or not set, yes or no) and as such is in a format that does not require conversion. In fact, the most common digital inputs are simple *contact–closure* devices, which provide a 0 or 1 to the computer depending on whether the associated process output relay is open or closed—indicating the state of the corresponding piece of equipment being monitored. Some examples are motor on–off, alarmed or normal, valve open–closed, gate closed–not closed (based on the position of a closed limit switch), and gate open–not open (based on the position of an open limit switch).

Signals can indicate normal operational events as status changes or abnormal events as alarms. Alarms may be a contact change of state (an intrusion alarm, for example) or a process variable exceeding acceptable limits (as with a high-temperature alarm). The digital input sending device thus notifies the computer of a status change through the I/O interface.

Digital outputs are switches operated by the computer. They can be used to turn equipment on or off, or to initiate sequential actions in the process. Digital outputs may also illuminate alarm panel windows or indicator lights.

Pulse data. Some sensors produce binary pulse trains encoding the sensed variable's magnitude as a function of the pulse rate. The pulses can be counted in the processor or the interface over a fixed time to produce a very accurate estimate of the amplitude of the sensed variable. An example is a kilowatt-hour measurement. Each time a wattmeter makes a complete rotation, a pulse is generated, indicating a fixed quantity of electrical energy. The computer detects and counts the frequency and duration of the pulses. The cumulative energy is logged or related to time to calculate the rate of electric power consumption.

Pulse outputs, consisting of a continuous train of proper size pulses over a selectable period, can be used to drive stepper motors of final control elements clockwise or counterclockwise to adjust to changes in set points. By turning the pulse generator on for a certain length of time, the computer can send any desired number of pulses to the process. When no pulses are being sent, the stepper motor does not move and the set point cannot change; hence, in the case of computer failure, the last set point will remain.

Controllers

Controllers are devices that take inputs from one or several process variables, apply a logical or mathematical function, and send the resulting output to a final control device. A variety of devices meet this definition, such as single-loop controllers, multiple-loop controllers, and programmable logic controllers (PLCs).

Programmable logic controllers. PLCs were originally designed to replace hardwired electromechanical relay circuits and reduce the expense of rewiring whenever sequential operations were changed. The origin of the PLC can be traced to the automotive industry where rapid sequential control using large banks of relays was common. According to the National Electrical Manufacturers Association a PLC is

> A digitally operating electrical apparatus that uses a programmable memory for the internal storage of instructions for implementing specific functions such as logic, sequencing, timing, counting and arithmetic to control, via digital or analog input/output modules, various types of machines or processes.

Therefore, PLCs are limited-function digital computers. PLCs originally were designed with a programming interface based on the relay ladder logic familiar to control engineers and technicians (see chapter 2). Current models also use function blocks and structured text as well as the more traditional ladder logic. They are designed to withstand wide variations in ambient temperatures, electrical noise, and vibration. As a result of competition and technological advances in the computer field, PLCs have overcome some of their original functional shortcomings. Better operator interface, easier program setup, improved arithmetic function, and local area networking capabilities can be found in the newer PLCs.

Distributed control units (DCUs). The term *distributed control unit* had its origin in the continuous process industry. Chemical and petrochemical plants had the need for process controllers imputing many analog variables (the PLC traditionally handled discrete variables). The DCU was originally a remote device designed to process complex analog control loops very rapidly. DCUs are usually not as rugged as PLCs and, therefore, require more environmental controls to protect them from heat and radio frequency interference. As with the PLC, the DCU has evolved to take on many of the features of its discrete counterpart.

Remote terminal units (RTUs). PLCs and DCUs are normally factory- or plant-floor devices that can be easily equipped with environmental controls. Because sensors and controllers had to be placed in harsh environments and communicate from off-site (remote) outdoor locations, the RTU was developed (see chapter 7). RTUs were originally developed as nonintelligent I/O devices, communicating via radio and telephone links. Newer RTUs incorporate full-function industrialized computers as their main processors and are essentially robust DCUs. Modern RTUs can provide full control and logical operations as well as operator interface.

Smart field devices. Field devices traditionally communicated with the DCU, PLC, and RTU using digital or analog (4–20 mA) inputs. These signals were then digitized by the I/O processor and sent to the CPU on a proprietary data channel. With smart field devices, a small microprocessor is placed directly within the end element (for example, sensor, valve, or pump) to act as the I/O processor. Field devices may now directly and digitally interface with the various computers in the system. As the I/O becomes more intelligent, logical control and proportional integral derivative controls are being added into the end elements. Digital fieldbus standards will allow direct digital communications between the sensors and control elements. The movement of the microprocessor to the field element has resulted in additional capabilities beyond normal I/O functions. Smart field devices can often be calibrated remotely. New international standards require device identification and diagnostics to be stored within the electronics of the smart device.

Software

Software is the collection of programs, routines, and instructions that controls the hardware of a digital control system, directing it to perform certain tasks. The two general types of software are operating system (OS) software and application software.

Operating system. An OS is a set of programs that controls the execution of applications programs and provides the interface between the applications programs and the hardware of the control system. An OS is normally supplied and packaged with the hardware and purchased as an integral part of the equipment. It manages system resources and does internal digital housekeeping.

An OS usually consists of three parts: a kernel or nucleus, a command interpreter, and the I/O and peripheral device drivers (see Figure 10-2). The kernel supplies the interface between the computer's hardware and software. The command interpreter allows a user to give commands to the OS. The command interpreter also makes system calls to the kernel for access to an I/O device or mass memory. I/O and peripheral device drivers control specific hardware peripheral devices, such as printers.

A real-time OS adequate for today's control systems includes real-time multitasking capability, an intertask messaging facility, and a disk file system. It can schedule tasks on the basis of external events that are signaled to the computer by interrupts and is designed to provide fast response to interrupt requests. Furthermore, a real-time OS is designed to prevent any task from monopolizing the processor. This prevents the computer from getting hung up and missing an important task. Its multitasking capability allows for the use of very complex real-time software. It also should supply prewritten and debugged software facilities, such as interrupt handlers, data transfer functions, and real-time clocks.

Application software. Application software includes programs developed to perform specific tasks. In digital control systems, application software enables a computer to monitor and control a process, such as operation of a water treatment plant or distribution system. Many standard software packages are available from system integrators and control system vendors that meet the needs of water systems. These standard packages offer many advantages to a utility considering implementing a control system, including the following:

- While the cost of hardware may decrease, software development is highly technical and labor intensive. The cost of developing a standard package can be borne by all purchasers, reducing the cost to each.

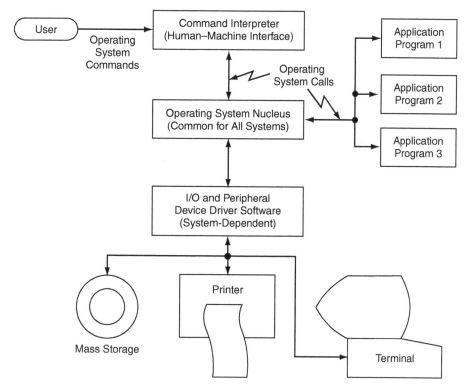

Figure 10-2 Operating system

- Vendors of packaged software generally have research and development groups improving and enhancing software. Upgrades are often available to current customers at a reasonable cost.

- Standard software is easier and less costly to maintain than custom software. Software support is often available from the vendor, alleviating the problems of losing the highly specialized people who maintain custom software.

- In a standardized system, software bugs are generally fewer than with custom software.

Proprietary application programs normally provide the tools necessary to build and maintain a database, write and modify control programs, display trends of process variables, build graphic displays, annunciate alarms, print reports, communicate with process I/O, and perform system diagnostics.

Even when packaged application software is used, a certain number of enhancements to the standard software will likely be required to account for the unique aspects of each application. The developer of these modifications could be the user's engineer or consultant, the vendor's or system integrator's engineer, an independent systems house, or a combination of these.

Also specific to the application is system configuration, which encompasses building the database, control programs, and operator interface displays. Application software configuration is best defined by the user or their engineering consultant and may be modified to suit the intended purpose of the control system. Specifically, water control systems will require the following data to be configured:

- *Point definition database.* Defines all process I/O that the control system will monitor and control. Typically, these will be digital inputs and outputs, analog inputs and outputs, and pulse counters. The definition will normally assign to each point a unique tag name, English description, and process I/O location. Analog points may also need measurement ranges, engineering units, and alarm limits entered.

- *Data display conventions.* Defines how the operator's console will display the value or status of each defined point in a standard, easily recognized way.

- *Graphics.* Creates and displays graphic representations of the process and plant layout, complete with dynamic real-time values and status from the database. Plant layout, flow path rates, and equipment operation are easily visualized in full-color displays.

- *Trends.* Defines the process parameters that will be displayed in a format that compares values over time, similar to the function of a strip-chart recorder. Trend displays allow operators to easily identify changes of state or value, analyze anomalies, and troubleshoot problems with the process. They are also useful to maintain, and display a historical record of process or equipment statuses.

- *Reports.* Defines how historical data will be incorporated into standard report forms. Daily, weekly, monthly, and yearly reports may be generated from stored data and structured suitably for specific needs.

- *Control.* Defines the proper control algorithms to analyze input data, perform logical (Boolean) or calculated (arithmetic) functions, and send appropriate outputs.

- *Security.* Defines what access to computer functions will be restricted to certain personnel. Control of security and level of privilege is generally maintained using passwords.

- *Historical data storage.* Defines which data will be stored for future retrieval and analysis. Data are generally stored in mass memory as they are compiled and offloaded to removable magnetic or optical media for permanent storage.

COMMUNICATION SYSTEMS

Control communications have evolved along with digital technology. In the early 1960s, analog control on a loop-by-loop basis was the norm, and operators adjusted the process manually. The first applications of digital technology replaced the analog controllers with digital equipment. Further benefits of computer technology greatly improved process optimization and reduced manual operation.

A computer can communicate with equipment by direct wiring—as with an analog controller—with a wire pair connected to the computer for each monitoring or control signal. Even within the confines of a water treatment facility, direct wiring can consist of miles of cable.

A more modern approach is to connect portions of the process to the control system using I/O subsystems where the equipment is located. These I/O subsystems are called *remote multiplexers* (muxes). A remote multiplexer can accept many direct-wired signals. At any given location, hundreds of signals may be connected to a remote multiplexer over relatively short runs.

Multiplexing is effective because of line sharing—the ability of the multiplexer to deliver to a computer sequentially scanned values of the monitored signals over a common data communication cable. Scanning involves the following steps: the instantaneous value of each variable is read, converted from an electric current to a coded digital value, and transmitted to the computer. The scanning sequence is predefined, so that the central computer recognizes the identity of each coded value on arrival, then stores that value in a reserved location in its database in memory. Each multiplexer is sequentially polled on a cyclic basis, to scan and deliver its complement of signals to the computer until inputs from all scanned points have been stored as latest values. Until the next scanning cycle begins, the stored values remain to represent the most recent system status for data processing, displays, control procedures, and for historical data storage access for logs and reports. In a similar but reversed method, control signals are periodically delivered from the computer back to the process via multiplexers.

As effective as remote multiplexing is in controlling a local process, its benefits are multiplied as the network grows in scope. As the number of remote sites and the distances from those sites to the central computer grow, cost effectiveness and efficiency increase. Many methods for communicating this data from the muxes to the DCUs or PLCs and between DCUs, PLCs, and OISs are available. More data require more speed, and long distances make speed expensive.

Digital Communication Concepts

The following paragraphs discuss the main concepts of digital communication.

Layers of communications. Depending on the system size, several layers of communications may be required, as shown in Figure 10-3. Each layer involves different communication needs and environments. Process level communications bring data from the field device to the controller. Information shared between controllers form the basis of a local area network (LAN). Linking several LANs forms a wide area network (WAN).

Standards. A variety of different standards and proprietary systems cover various parts of network communications. Because of the development of so many different standards, each only addressing part of the overall requirements, an effort was launched to standardize digital communication systems. In 1982 the International Standards Organization (ISO) proposed a unified method for communicating with a computer in the form of multiple independent layers. This model for protocols (rules and conventions that govern the transmission of data) is named the Reference Model for Open System Interconnection. The ISO model has defined seven layers of protocols in the communication network as shown in Figure 10-4. Each layer transmits a message from one layer to the next.

Many electric systems comply with the Electronic Industries Association (EIA) standards. EIA standards were originally developed for teletype communications on public telephone networks. For example, RS-232C defines the electrical characteristics of the interface between a data terminal or computer and a 25-conductor communication cable. RS-232C further defines voltages, data transmission speeds, and distances.

The RS-232C standard was written primarily for a communication link with a single transmitter and single receiver over a short distance (50 ft [15 m]). To overcome some of these limits, other standards were developed. RS-423A expanded

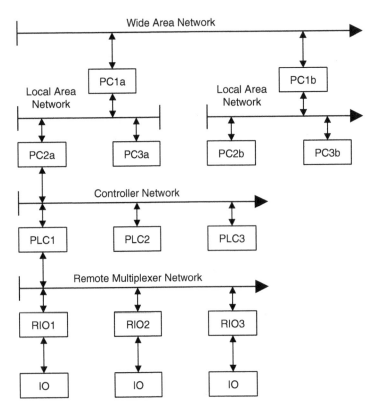

Figure 10-3 Layers of communications, LAN, WAN

ISO Model	
Application	Layer 7
Presentation	Layer 6
Session	Layer 5
Transport	Layer 4
Network	Layer 3
Data Link	Layer 2
Physical	Layer 1

Figure 10-4 Reference model for open system interconnection

the capabilities to 10 receivers; RS-422A incorporated differential devices to achieve better noise resistance; RS-485 expanded the system to 32 devices at distances up to 4,000 ft (1,200 m). A listing of parameters for commonly used EIA standards is given in Table 10-1.

These standards only define physical requirements. To transmit meaningful data, a protocol or message structure must also be defined.

To continuously move large quantities of data between computer systems, a very strict set of rules must be followed. A wide variety of international standards evolved to cover the requirements of LANs and WANs.

Table 10-1 EIA standards

Standards Characteristics	EIA-232E	EIA-422A	RS-485
Allowed number of transmitters and receivers per data line	1 Tx, 1 Rx	1 Tx, 10 Rx	32 Tx, 32 Rx
Maximum cable length	50 Feet	4,000 Feet	4,000 Feet
Maximum data rate	20 kbps	10 Mbps	10 Mbps
Minimum voltage	±5 Volts	±2 Volts	±1.5 Volts
Maximum voltage	±15 Volts	±5 Volts	±5 Volts

For example, the Institute of Electrical and Electronics Engineers (IEEE) established a series of standards covering various aspects of data communication:

- IEEE 802 covers large networks and multiple path connections.
- IEEE 802.2 covers the data link layer.
- IEEE 802.3 covers multiple action/collision detection media access control, the basis of Ethernet.
- IEEE 802.4 covers token-bus media access control.
- IEEE 802.5 covers token-ring media access control.

Remote communications. Distant communications are commonly handled by either radio or telephone systems. Whether a telephone or radio is used, variables measured in the field (or information sent to the field) are converted into signals suitable for the transmitting system. A modem converts digital computer data into a digital form suitable for the radio or telephone system. Currently, a wide variety of modems are available, which can be used on dial-up telephones, leased lines, fiber optics, or radio systems.

There are many modem standards. These standards have evolved as electronics became more capable of handling higher speeds. In addition to speed, there are many standards that address various data compression methods and error-control techniques. As the quality of electronics and telephone networks improves, newer standards will be developed.

Networks. Achieving distributed yet unified control systems depends on a communications network in which a number of distributed process control units may be linked to a central computer, an OIS, and to each other. A *link* is a communication channel that connects two nodes. The link may be a twisted-pair cable, telephone line, fiber-optics cable, coaxial cable, microwave link, or other type. There are nearly as many network arrangements as control systems, but most are one of the following types or a combination (see Figure 10-5):

- *Star.* The star network consists of a central node (often in a distributed process controller or the central computer itself) connected to peripheral nodes by individual communications links, one link to each peripheral node.

- *Ring.* In a ring configuration, each node is connected to two other nodes (usually two adjacent nodes). Messages received by one node are passed along, or relayed, to the destination node. Any node can normally initiate a message, and because the messages are passed through intermediate nodes that act as repeaters, communications over long distances are possible.

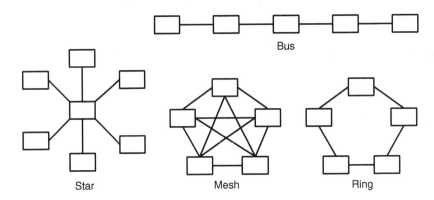

Figure 10-5 Networks

- *Multidrop or bus.* In a multidrop network, a single link is connected to all the nodes. This extended link is also known as a *data bus* or *data highway*. The extended link is a very economical network architecture and is commonly used in process control environments. Any two nodes can communicate, but generally only two at a time. Increased communication speed available with separate dedicated links (peer-to-peer) and improved methods of avoiding message contention have reduced the impact of this shortcoming, but obviously, the more nodes connected to the data highway contending for communication time, the slower the network becomes.

- *Multiconnected.* In a multiconnected network, each node that needs to communicate with another node is connected by a separate link. This type of network has high reliability in that if one link fails, other paths might be available to permit communication via a third node. This structure requires higher level communication equipment to insure proper routing.

In many complex communication network configurations, the nodes communicate over the network with a central computer or a dedicated communication controller directing traffic. To ensure that all information is correctly routed, every data message contains an address that uniquely identifies its destination among the distributed units. A buffer in the destination node quickly captures the burst of information, freeing the data network to turn immediately to the next priority message. The buffer then decodes the message for the node.

Continuous polling. Continuous polling describes a method of data exchange between the master and remote stations. The master stations transmit a specific address over the communication network. Because this is a common system, all remote devices will receive this information, but only the one whose address matches will respond. Once the information is passed between the master and this specific address, communications are stopped, and the next addressed is polled. This process is continued until all remotes have communicated with the master. Once completed, the cycle is started again.

Reports by exception. The sequential polling of many remote modules with large I/O connections is time consuming. In water systems, many of the data remain unchanged from one scan to the next. To shorten scan time, the remote terminal units can be programmed to respond only to changes that have occurred since its last scan. This is referred to as *report by exception*.

High-speed networks. Different areas of the network operate at different rates of speed. Process variables in water systems normally can be transferred at relatively slow rates. Higher rates in the network are required to transfer large data blocks, such as large files, graphics, or reports. Because of the number of data involved, a much higher rate of data transfer is required. However, the relationship between network speed and data transfer or throughput must be understood. Network communications include both data and security information. High-security applications transmit more security-related information than data. High-speed systems that require multiple *handshakes* (see the section, Error detection/error correction) or error checks may have relatively slow throughputs. Total speed also depends on the speed of all connected devices. A fast data highway with low-speed gateways is limited to the slowest component, which becomes a *bottleneck*.

Another factor which slows down network throughput is retransmissions. These occur whenever a message is not received correctly. This may be caused by broken equipment, adverse line conditions (like the affect of weather on radios), or heavy network traffic causing message collisions. Message collisions result from two or more nodes trying to speak at once. This is similar to two people trying to talk at the same time in a meeting. Controlling collisions has a significant impact on network throughput.

Three common LANs are Ethernet, Token-Ring, and ARCNet. Simply rating them by speed (ARCNet: 2.5 megabits per second; Token-Ring: 4 megabits per second; and Ethernet: 10 megabits per second) can lead to the wrong conclusions regarding throughput. Depending on network loading, equipment, and routing, the actual data throughput may vary greatly. For example, heavy traffic can slow down Ethernet more than the others because of Ethernet's poor collision control. Under these conditions, the resultant performance (throughput) may be much worse than the others even though Ethernet's speed is much higher.

Error detection/error correction. The movement of data in a communication system involves many shared components and exposes the communication system to possible effects of the environment. Outside electrical forces, if unchecked, can interfere with a digital transmission system. During data movement, some of the information will probably not reach its final destination in an error-free state. To minimize the occurrence of bad data, many systems provide several levels of security. The first level typically implements a *handshake,* which is the acknowledgment of the receipt of data sent. A second method of error detection is usually encoded into the message structure to verify the correct amount or state of bits have been received. The third level of security uses a mechanism to resend data that have been determined to be corrupt or incomplete that may have passed through the lower levels. No system is error free. By applying proper security, the vast majority of errors can be controlled.

Connectivity. Connectivity is used to describe a method of interconnecting computers of the same manufacture. The current definition defines the interconnection between computer systems made by many different manufacturers. Connectivity covers signaling, cabling, and software. Many of the products of connectivity cover the standards produced by the ISO while others use proprietary software to attempt to achieve the same results. The complexity of connectivity is a function of the number of users and the type of data transfers required. A small system with few users and minimal interface will require a much simpler structure than one with 20 to 100 or more users.

Communication media. The type of media used to transmit the digital information is critical to system performance. The media must be capable of conveying the data at a desired speed and distance at an acceptable cost. The ability to connect, test, and modify are important factors in media usage. Chapter 7, Telemetry, discusses communication media in more detail.

APPLICATIONS AND SITE PLANNING

The first control systems had, essentially, distributed control. That is, individual elements of the process were physically separate, and the indicators (level, pressure, temperature, status) and control elements (valves, switches, pumps) were largely local. Operators for these process elements were stationed locally. With the advent of pneumatic and electronic signal transmission systems, centralized control stations with indicators, control devices, and recorders became the accepted method of process control. As remote multiplexing became available, system communications became more complex.

Centralized automation became practical as mainframe computers or minicomputers became affordable alternatives to direct operator control. Most of the effects of these changes were beneficial, but they were not achieved without some trade-offs. One of these is the dependency on the central computer hardware and software as well as communication links between the computer and the remote multiplexers at the process element sites. Fortunately for process control system operators, technological developments in the computer field did not stop with minicomputers.

Large-scale integrated circuits have made it economically feasible to locate microprocessors and microcomputers around a plant or process at individual control elements. These microcomputers are located not only in individual instruments, but in programmable logic controllers and in remote terminal units. These devices cooperate with remote multiplexers to facilitate communications and perform local monitoring and control. In so doing, they alleviate many of the previously mentioned problems that had plagued many centralized automation systems. Specifically, microcomputers:

- Reduce reliance on the central computer. If the central computer fails, local control can continue, and updates of the central computer may resume when its operability is restored.

- Limit communications over the network to essentials. Local control (in which set points are derived and maintained locally) can be accomplished by the distributed processor according to programmed routines without interrupting the central computer.

- Decrease dependence on the communications link to the central computer.

- Reduce the overall computational load of a central computer, limiting the size (computational power) of the requisite central unit. Some modern digital control systems operate without a central computer altogether—all process control being located remotely, with reports, logs, and historical data handled by a separate dedicated computer.

A practical application of this distributed control would be to operate a pump station at a site separate from the control center. Monitoring and control applications are programmed as software in the distributed processing unit located at the pump station. The processor (along with all the other processors on the network) proceeds independently through the cycle of data scanning, sampling, testing, evaluating, and delivering control outputs as necessary. On a periodic basis, the central computer interrogates each distributed processor for an update of status changes, alarms, and latest values of scanned variables. From the data received, the central computer assembles and presents displays, printed logs, and historical data storage. Using the operator's console with its keyboard and VDU display, the system operator can execute control commands or make changes to the central or local database and control programs. In many cases, the differences between a centralized and a distributed digital control system will be unrecognized by the operator.

The majority of computer applications in water systems acquire information. Computers in small plants may be dedicated to monitoring only. The following list covers some of the items that a typical water system can monitor or control with digital systems.

- Remote pumps, wells, reservoirs, or tanks
  - Monitor—levels, flows, pressure, pump running, valve position, power, and security
  - Control—pump operation and valve operation
- In-plant
  - Monitor—raw and finished flows, turbidity, levels, pH, motor running, motor current, failure alarms, chlorine alarms, chemical usage, and valve position
  - Control—influent rate control, chemical feed system, chemical mix system, pump control, valve control, sludge blowdown, dewatering and conveyance filter backwash control, chlorine control, and distribution control

Site Planning

Careful planning of the control room and remote sites is not merely providing space for control system equipment, but encompasses details that can determine success or failure of control system implementation.

Control room. Vital issues that should be addressed in control room site planning are discussed in the following paragraphs.

Control room layout should take into account not only which equipment needs to fit in the allotted space, but also when and how many operations and maintenance personnel will need access to the equipment. Frequency and duration of that access should also be considered. If a 24-hour attended operation is planned, ergonomic needs must be carefully considered. Comfort, health, and safety of operations personnel are vital to ensure maintenance of high morale and efficiency.

The operations console should be designed to facilitate proper access to and operation of the equipment. Adequate counter or desk space, storage of and access to manuals, number of concurrent operators or users, ergonomic issues such as height of screens and adjustability of keyboard placement, necessity of voice and video communication equipment, and appearance are all aspects that merit consideration.

Heating, ventilation, and air conditioning (HVAC) loads should be determined and adequate HVAC provided. The comfort and safety of operations personnel as well as the reliability of equipment must be considered. The criticality of the equipment and use of the control room should be considered in determining the need for redundant HVAC equipment.

Fire detection and protection must be planned. Standard fire suppression systems (sprinklers) have limited applicability in the control room because water may be as destructive to equipment as smoke or fire, and a false alarm could trigger a catastrophe. Suppression systems using fire retardant gases are available. Gases that do not damage the ozone layer are replacing Halon in these systems. Communication equipment must be protected as well as computer equipment against fire, even if it isn't located within the control room.

An uninterruptible power supply (UPS) is an essential component of a digital control system. Power surges, noise, transients, and interruptions of varying duration are potentially devastating to a 24-hour real-time control system. A UPS isolates the computer equipment from fluctuations, at the same time providing short-term standby power during power failures, providing clean alternating current exclusively for computer hardware. An alternate source of power that lasts longer than the battery backup provided by the UPS should be installed, usually an auxiliary generator, sized to support all vital equipment. This generator, powered by an alternative energy source, can be automatically started during power failures that exceed a certain duration. Attaching generator power leads to the UPS gives the dual advantage of isolation from power irregularities present in the auxiliary power supply as well as allowing the UPS batteries to provide a smooth transition from one source to the other.

Remote site. Environmental factors at control and monitoring sites in a typical water utility application are variable, and especially with SCADA systems, where some sites may be quite remote from the control center in undeveloped areas. Access, security, temperature, moisture and humidity, power supply, and communications are all factors that may affect operational reliability and must be considered if a system is to meet specified operational availability expectations.

Site planning for remote sites generally requires more compromises than for control rooms. Because of the relatively large number of separate sites, protective measures that are practical when applied to a single (control room) site are not cost effective. Fortunately, failure of a single remote site is seldom fatal to the overall operation. In cases where it would be, appropriate additional protection should be provided.

Adequate space for equipment installation and maintenance should be included in the plan. Potentially damaging environmental factors, such as rain, snow, humidity, dust, high or low temperature, should be considered in the design.

The availability, reliability, and condition of electric power at remote sites are major concerns of SCADA systems users. Power surges (especially those induced by lightning) may disrupt communications or even destroy equipment. Manufacturers' requirements for grounding, power conditioning, and surge protection must be heeded to minimize costly damage and prevent operational disruptions (see chapter 2). Often some sort of battery backup power is used to keep remote site monitoring and communication online in case of power failure. Solar panels are often used to power RTUs at sites remote from the electric utility power grid.

TECHNOLOGY TRENDS

Water system operators are besieged from many directions. Governmental regulations are requiring better control and more reports on operations. Environmental concerns are promoting the monitoring of a growing list of potential health hazards. Funding for many plants is shrinking. Computerized automation is a tool which can help operators deal more effectively with these new requirements.

Experience with many automated systems in the past has resulted in a cautious approach to the implementation of any computerized system. Many of the early systems did not perform as required. On the positive side, many of the early problems have been overcome by newer and more reliable electronics. Operator interfaces in the newer systems no longer require the learning of a new language.

Technology trends greatly favor the water plant operator. Smarter electronics will not only pass process variables but also keep track of equipment maintenance. Predictive maintenance programs will become commonplace. Emergency repairs will be minimized as equipment can be scheduled for repair in a predictable manner. Expert systems will assist the plant operator in modifying an operation to meet changes in systems requirements. Computer networking will give data to management in a form that will better guide their decisions. Properly applied automation in water systems will greatly increase operational efficiency.

REFERENCES

American Water Works Association Research Foundation/Japan Water Works Association. 1994. *Instrumentation and Computer Integration of Water Utility Operations.* Denver, Colo.: AWWA/AwwaRF.

Anderson, N. A. 1980. *Instrumentation for Process Measurement and Control.* Radnor, Pa.: Chilton Company.

Batten, G. L. 1988. *Programmable Controllers.* Blue Ridge Summit, Pa.: Tab Books, Inc.

Carr, J. J. 1988. *Data Acquisition and Control.* Blue Ridge Summit, Pa.: Tab Books, Inc.

Hodson, R. F., and A. Kandel. 1991. *Real-Time Expert Systems Computer Architecture.* Boca Raton, Fla.: CRC Press, Inc.

Lawrence, P. D., and K. Mauch. 1987. *Real-Time Microcomputer System Design.* New York: McGraw-Hill, Inc.

Lenk, J. D. 1984. *Handbook of Microcomputer-Based Instrumentation and Controls.* Englewood Cliffs, N.J.: Prentice-Hall, Inc.

Money, S. A. 1986. *Microprocessors in Instrumentation and Control.* New York: McGraw-Hill, Inc.

Rosenof, H. P., and A. Ghosh. 1987. *Batch Process Automation—Theory and Practice.* New York: Van Nostrand Rheinhold.

Skrenter, R. G. 1989. *Instrumentation.* Chelsea, Mich.: Lewis Publishers, Inc.

Skrokov, M. R. 1980. *Mini- and Microcomputer Control in Industrial Processes.* New York: Van Nostrand Rheinhold.

Stoecker, W. F., and P. A. Stoecker. 1989. *Microcomputer Control of Thermal and Mechanical Systems.* New York: Van Nostrand Rheinhold.

This page intentionally blank.

AWWA MANUAL M2

Chapter **11**

Instrument Diagrams

A standard instrument diagram or process and instrument diagram (P&ID) is frequently used to define the instruments and functions of a control system in the water utility industry. By standardizing these diagrams, the users, consultants, and equipment manufacturers will be able to understand one another. For instance, contractors and equipment manufacturers will be able to easily interpret specifications written by utilities and their consultants.

For this reason, the Automation and Instrumentation Committee has recommended that the Instrument Society of America (ISA) standard notation for instrument diagramming be used (see Table 11-1). Because of the limited scope of this manual, only the basics of the ISA standard notation are presented. The complete Standards and Recommended Practices for Instrumentation and Control are available from the ISA.

When P&ID standards were initially developed, each symbol represented a specific piece of hardware that was required to perform a specific function. Now, one hardware component, such as a programmable controller or computer, provides many functions in an instrument loop. For example, a single programmable controller or computer may be used to extract the square root of the differential pressure signals from five loops while performing signal comparisons for high- and low-level alarms in ten other loops.

Even though a single piece of hardware may be used to perform multiple functions, all of the functions of the hardware should be shown on the P&ID to maintain the integrity of the P&ID as a communication tool. Each instrument loop can be easily identified with a series of letters that define the functions of the instrument loop. The letters in Table 11-1 are used in conjunction with the general instrument or function symbols shown in Figure 11-1. The function designations for relays shown in Figure 11-2 are typically used to add supplemental information to the symbols shown in Figure 11-1. Standard instrument line symbols are shown in Figure 11-3. The summary of special abbreviations shown in Table 11-2 may also be used to define further the P&ID. See example P&ID in Figure 11-4. Please note that the intent of the example P&ID shown is to depict instrument symbols, not to define a specific process.

Table 11-1 ISA instruments

	First Letter		Succeeding Letters		
Ltr.	Measured or Initiating Variable	Modifier	Readout or Passive Function	Output Function	Modifier
A	Analysis		Alarm		
B	Burner, Combustion		User's Choice	User's Choice	User's Choice
C	User's Choice			Control	
D	User's Choice	Differential			
E	Voltage		Sensor (Primary Element)		
F	Flow Rate	Ratio (Fraction)			
G	User's Choice		Glass, Viewing Device		
H	Hand				High
I	Current (Electrical)		Indicate		
J	Power	Scan			
K	Time, Time Schedule	Time Rate of Change		Control Station	
L	Level		Light		Low
M	User's Choice	Momentary			Middle, Intermediate
N	User's Choice		User's Choice	User's Choice	User's Choice
O	User's Choice		Orifice, Restriction		
P	Pressure, Vacuum		Point (Test) Connection		
Q	Quantity	Integrate, Totalize			
R	Radiation		Record		
S	Speed, Frequency	Safety		Switch	
T	Temperature			Transmit	
U	Multivariable		Multifunction	Multifunction	Multifunction
V	Vibration, Mechanical Analysis			Valve, Damper, Louver	
W	Weight, Force		Well		
X	Unclassified	X Axis	Unclassified	Unclassified	Unclassified
Y	Event, State or Presence	Y Axis		Relay, Compute, Convert	
Z	Position	Z Axis		Driver, Actuator, Unclassified Final Control Element	

INSTRUMENT DIAGRAMS 201

	Primary Location Normally Accessible to Operator‡	Field Mounted	Auxiliary Location Normally Accessible to Operator‡
Discrete Instruments	1 ⊖ * IP1†	2 ○	3 ⊖
Shared Display, Shared Control	4	5	6
Computer Function	7	8	9
Programmable Logic Control	10	11	12

*Symbol size may vary according to the user's needs and the type of document. A suggested square and circle size for large diagrams is shown above. Consistency is recommended.

†Abbreviations of the user's choice such as IP1 (Instrument Panel #1), IC2 (Instrument Console #2), CC3 (Computer Console #3), etc., may be used when it is necessary to specify instrument or function location.

‡Normally inaccessible or behind-the-panel devices or functions may be depicted by using the same symbols but with dashed horizontal bars, i.e.,

Figure 11-1   General instrument or function symbols

The function designations associated with relays may be used as follows, individually or in combinations. The use of a box enclosing a symbol is optional; the box is intended to avoid confusion by setting off the symbol from other markings on a diagram.

SYMBOL	FUNCTION
1. 1 – 0 or – OFF	Automatically connect, disconnect, or transfer one or more circuits provided that this is not the first such device in a location.
2. Σ or Add	Add or totalize (add or subtract)†
3. Δ or Diff.	Subtract†
4. ± ÷ ⊟	Bias*
5. Avg.	Average
6. % or 1:3 or 2:1 (typical)	Gain or attenuate (input:output)*
7. ⊠	Multiply†
8. ⊞	Divide†
9. √ or Sq. Rt.	Extract square root
10. x^n or $x^{1/n}$	Raise to power
11. f(x)	Characterize
12. 1:1	Boost
13. ▷ or Highest (Measured Variable)	High-select. Select highest (higher) measured variable (not signal, unless so noted).
14. ◁ or Lowest (Measured Variable)	Low-select. Select lowest (lower) measured variable (not signal, unless so noted).
15. Rev.	Reverse
16.	Convert
a. E/P or P/I (typical)	For input/output sequences of the following:

Designation	Signal
E	Voltage
H	Hydraulic
I	Current (electrical)
O	Electromagnetic or sonic
P	Pneumatic
R	Resistance (electrical)

b. A/D or D/A	For input/output sequences of the following:

Designation	Signal
A	Analog
D	Digital

17. ∫	Integrate (time integral)
18. D or d/dt	Derivative or rate
19. I/D	Inverse derivative
20. As required	Unclassified

*Used for single-input relay.
†Used for relay with two or more inputs.

Figure 11-2 Function designations for relays

ALL LINES TO BE FINE IN RELATION TO PROCESS PIPING LINES

1. Instrument Supply* or Connection to Process
2. Undefined Signal
3. Pneumatic Signal†
4. Electric Signal
5. Hydraulic Signal
6. Capillary Tube
7. Electromagnetic or Sonic Signal‡ (Guided)
8. Electromagnetic or Sonic Signal‡ (Not Guided)
9. Internal System Link (Software or Data Link)
10. Mechanical Link

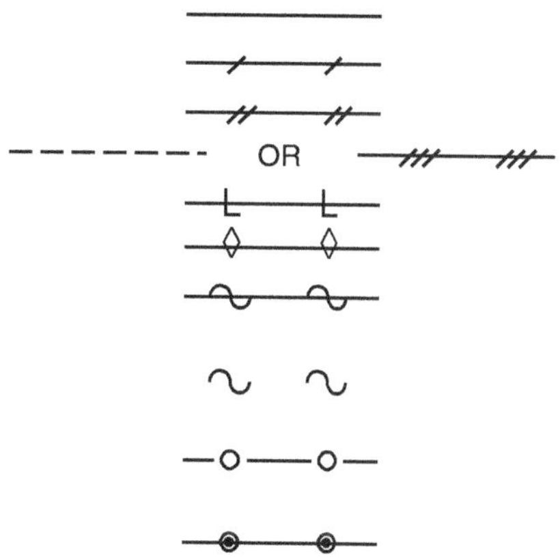

Optional Binary (On–Off) Symbols

11. Pneumatic Binary Signal
12. Electrical Binary Signal

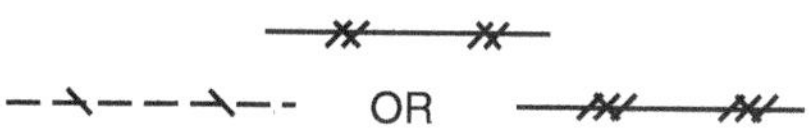

NOTE: "OR" means user's choice. Consistency is recommended.

*The following abbreviations are suggested to denote the types of power supply. These designations may also be applied to purge fluid supplies.

  AS – Air Supply
    IA – Instrument Air ⎫
    PA – Plant Air   ⎬ Options
  HS – Hydraulic Supply
  NS – Nitrogen Supply
  SS – Steam Supply
  WS – Water Supply
  ES – Electric Supply
  GS – Gas Supply

The supply level may be added to the instrument supply line, e.g., AS-100, a 100-psig air supply; ES-24DC, a 24-volt direct current power supply.

†The pneumatic signal symbol applies to a signal using any gas as the signal medium. If a gas other than air is used, the gas may be identified by a note on a signal symbol or otherwise.

‡Electromagnetic phenomena include heat, radio waves, nuclear radiation, and light.

Figure 11-3 Standard instrument line symbols

Table 11-2 Summary of special abbreviations

Abbreviation	Meaning
A	analog signal
ADAPT.	adaptive control mode
AS	air supply
AVG.	average
C	patchboard or matrix board connection
D	derivative control mode
	digital signal
DIFF.	subtract
DIR.	direct-acting
E	voltage signal
ES	electric supply
FC	fail closed
FI	fail indeterminate
FL	fail locked
FO	fail open
GS	gas supply
H	hydraulic signal
HS	hydraulic supply
	current (electric) signal
I	interlock
	integral control mode
M	motor actuator
MAX.	maximizing control mode
MIN.	minimizing control mode
NS	nitrogen supply
O	electromagnetic or sonic signal
OPT.	optimizing control mode
P	pneumatic signal
	proportional control mode
	purge or flushing device
R	reset of fail-locked device
	resistance (signal)
REV.	reverse-acting
RTD.	resistance temperature detector
S	solenoid actuator
S.P.	set point
SQ. RT.	square root
SS	steam supply
T	trap
TEL.	telemetering signal
WS	water supply
X	multiply unclassified actuator

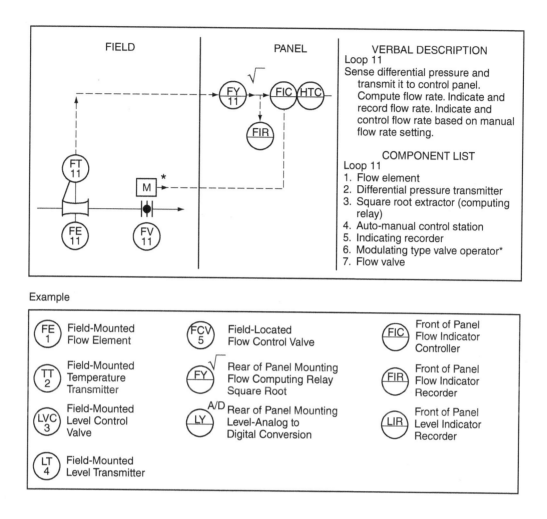

Figure 11-4   Example of PI&D loop description

Standard notations are enhanced with narrative descriptions and component lists as supplements to the P&ID. The notations should clarify designer's intent. The following paragraphs include the loop description and component list for Loop 1 shown in Figure 11-4:

Loop description:

- Measure and locally indicate raw water pressure.

- Transmit a proportional signal to the programmable controller.

- Control raw water pressure by modulating pressure control valve PCV-1.

- Alarm high and low pressure at the plant computer.

- Indicate raw water pressure at the plant computer.

Components:

- PIT-1—Pressure Indicating Transmitter
- PC/PS-1—Pressure Controller/Switch (Programmable Controller functions)
- PY-1 [I/P]—Pressure Relay (Current to Pneumatic converter)
- PCV-1 [FO]—Pressure Control Valve (Fail Open)
- PI-1—Pressure Indicator (Computer function)
- PA-1 [PAH PAL]—Pressure Alarm High, Pressure Alarm Low (Computer functions)

Glossary

(Note: *Italicized* words in the definitions are defined in this glossary.)

absolute pressure   *Gauge pressure* plus atmospheric pressure at reference sea level elevation.

AC   Alternating current. An electrical system designed to have *current* flow changing direction repeatedly on a regular cycle. The *frequency* of the direction changes is measured in *hertz,* which is equivalent to *cycles per second.* This is the opposite of *DC.*

accuracy   The extent to which the results of a calculation or the readings of an instrument approach the true values of the calculated or measured quantities, and are free from error.

ampere   The basic unit of electrical *current.* Equal to one *coulomb* per second. See *Ohm's law.*

amplification   The production of an output of greater magnitude than the input.

apparent power   The total *power* in an *AC* system. It is the *vector sum* of *true power* and *reactive power.*

attenuation   The production of an output of lower magnitude than the input.

automatic control   Maintenance of balanced conditions within a process without the intervention of humans.

automatic controller   A device that automatically executes the decision making necessary to determine the proper adjustment of a *final control element* in response to measured conditions of the process.

averaging Pitot flowmeter   A flowmeter that consists of a rod extending across a pipe with several interconnected upstream holes, which simulate an array of Pitot tubes across the pipe, and a downstream hole for the static pressure reference.

bellows gauge   A device for measuring pressure in which the pressure on a bellows, with the end plate attached to a spring, causes a measurable movement of the plate.

beta ratio   The ratio of the diameter of the constriction (or orifice) to the pipe diameter in a differential pressure-type flowmeter.

bias   A fixed amount that is always added or subtracted to a measurement. For example, if a tank level is measured as water depth by a level sensor but water surface elevation is needed for operations, then a bias equal to the elevation of the bottom of the tank can be added to the measurement of depth. This would result in the readout indicating the elevation of the water surface.

bluff body   The name given to the obstruction (body) in a *vortex-shedding meter* to enhance the generation of vortices in the flow around the body and downstream in the meter.

calibration   The determination, checking, or rectifying of the graduation of any instrument giving quantitative measurements.

capacitance   An electrical property that stores electrical *charge*. It is measured in *farads* and is analogous to water tanks in a hydraulic system. Capacitance has the effect of resisting changes in voltage. The ratio of the charge on one of the conductors of a capacitor (there being an equal and opposite charge on the other conductor) to the potential difference between the conductors.

cavitation The action resulting from forcing a stream flow to change direction in which reduced internal pressure causes dissolved gases to expand, creating negative pressure. Cavitation frequently causes pitting of the hydraulic structure affected.

charge The amount of excess electron (*negative* charge) or the amount of excess protons (*positive* charge) in a material or physical area. If the number of electrons and protons are equal, then the charge is zero.

circuit breaker A mechanical device for limiting *current* flow as well as providing a switch for opening or closing an electrical circuit. When the current exceeds the limiting value, the switch is automatically opened. When the condition causing the excessive current is fixed, the circuit breaker can be reset manually and continue normal operation. The method of measuring current can be magnetic or thermal or both. Circuit breakers are sized by the amount of current they will pass without opening the switch and by how many *poles* they have. Each pole measures the current and contains a switch for a single conductor (wire).

compound loop control Refers to control loops which use both *feedforward control* and *feedback control* to control the same *final control element*.

conductivity The property of a water sample to transmit electric *current* under a set of standard conditions. Usually expressed as microsiemens per centimeter (µS/cm).

conductors Materials which can be easily forced to carry *electricity* are called conductors.

control action The particular mathematical relationship between an *automatic controller's* input and output. Synonymous with *control mode*.

control loop The combination of all instruments and devices which directly affect the position of a *final control element*.

control mode Also called *control action* and refers to the mathematical relationship between the output and the input of an *automatic controller*.

control relay An electromechanical device used to switch one or more electrical control signals in response to a trigger signal. They contain a switch of one or more *poles* which is operated by an electromagnet. They are usually small because they are designed to carry control signals rather than *power*.

controlled variable The quantity or condition that is measured and controlled.

coulomb The basic unit of electrical *charge*. Equal to the collective charge of 6.25×10^{18} electrons (or protons).

current The speed of electrical *charge* movement. Measured in *amperes*. See *Ohm's law*.

cycles per second A unit for measuring the *frequency* of alternating current. Abbreviated as cps and equal to *hertz*.

DC Direct current. An electrical system designed to have *current* flow in one direction only. This is the opposite of *AC*.

dead time Any definite time period required for any process material to propagate or travel between two different locations. Dead time can occur in the measuring system, the controller, or any of the process streams.

derivative (action or control) The *control mode* which varies the controller's output in proportion to the derivative (rate of change) of the *controlled variable*. Also called *rate action or control*.

derivative factor The amount of *derivative action* used in a particular *control loop*.

deviation The difference between the *controlled variable* and the controller *set point*. Also known as *process error*.

diaphragm A thin, flexible partition (disc) supported at the edges, used to transmit pressure from one substance to another while keeping them from direct contact.

differential pressure The difference in pressure between two pressure sources, measured relative to one another.

differential pressure gauge An apparatus to measure pressure difference between two points in a system; it can be a pressured liquid column balanced by a pressured liquid reservoir, a formed metallic pressure element with opposing force, or an electrical–electronic gauge (such as strain, thermal conductivity, or ionization.)

discharge coefficient A coefficient representing the ratio of true flow (measured by a high *accuracy* device) to the indicated meter flow.

disconnect switch A hand-operated switch for isolating electrical *power* from a device to permit safe maintenance on the device. The switch must open all power conductors running to the load and is always lockable in the off (or open) position. This is not necessarily the same as an *LOS* switch, which is used to prevent operation of electrical equipment.

Doppler-effect flowmeter The change in the observed frequency of an acoustic or electromagnetic wave due to relative motion of source and observer.

Doppler ultrasonic flowmeter An instrument for determining the velocity of fluid flow from a Doppler shift of high-frequency sound waves reflected from particles or discontinuities in the flowing fluid.

drift Wandering of the *controlled variable* around its *set point*.

earth The primary electrical *ground* in a power system. All electrical *power* and lightning systems have conductors known as grounding conductors to carry errant *currents* safely into the earth where the *charge* can be safely dissipated.

electricity The movement of electrical *charge* from one location to another.

electromotive force The basic electrical force. Often abbreviated EMF. Measured in volts.

element Any single discrete part of a system.

EMF See *electromotive force*.

error Difference between actual and desired values. See also *process error*.

farad The unit of measurement for electrical *capacitance*.

fault A short circuit in an electrical system. Any unintentional connection between conductors in an electrical power system.

FCE See *final control element*.

feedback control A type of control that measures the output of a process and adjusts one or more inputs to cause the output to assume a desired value or *set point*.

feedforward control A type of control that measures one input of a *process* and adjusts one or more other inputs to cause the output to assume a desired value.

fiber optics The use of fine glass fibers for transmitting signals by means of optical light.

final control element The actual device used to adjust the process. Motor speed controllers, chemical feeders, and control valves are common examples.

fluorescent The property of substance to emit visible light.

frequency The number of times that *current* flows in a given direction in one second in an *AC* system. It is measured in *hertz* or *cycles per second*.

fuse A device for limiting the amount of electrical *current* in a circuit. When an excessive current flows, the interior conductor of the fuse melts due to the heat generated by the current and thus opens the circuit. Unlike a circuit breaker, a fuse cannot be reused; it must be replaced.

gauge pressure That pressure that is in excess of atmospheric pressure at the referenced elevation.

ground Any common point in an electrical circuit against which electrical properties are measured. The portion of an electrical power circuit that is connected to *earth*.

hazardous area Any area in which flammable vapors or dust particles are likely to be present. Electrical equipment used in such areas must be specially designed to prevent arcs from igniting the atmosphere.

henry The unit of measurement for electrical *inductance*.

hertz The unit of measure of the *frequency* of *alternating current*. Equal to the number of times the current flows in a given direction in each second. Also expressed as *cycles per second*. Abbreviated Hz.

hunting Periodic changes in the *controlled variable* from one side of the *set point* to another as the controller overcorrects first in one direction, then in the other.

impedance The complex ratio of electric voltage to *current*.

inductance An electrical property that stores electrical energy in a magnetic field. It is measured in *henrys* and is analogous to inertia in a hydraulic system. *Induction* has the effect of resisting changes in *current* flow.

induction The action of using a magnetic field to transfer energy into an electrical circuit.

induction motor Any *motor* which relies on magnetic induction to couple energy into the *rotor* to cause the necessary rotor *currents*.

insulators Those materials that can only be forced to carry *electricity* by applying a very strong force are called insulators.

integral (action or control) The *control mode* which varies the controller's output in proportion to the integral (sum over time) of the *process error* in the *controlled variable*. Also called *reset action or control*.

integral factor the amount of *integral action* used in a particular *control loop*.

kilo A measurement unit prefix used to indicate "times one thousand." For example, kilovolt = 1,000 volts; kiloamp = 1,000 amperes; etc.

lag Any combination of inertial factors that prevents a process from changing instantaneously.

laminar flow A nonturbulent flow regime in which the stream filaments glide along the pipe axially with essentially no transverse mixing. Also known as viscous or streamline flow. Streamline flow in a viscous fluid. Flow under conditions in which forces due to *viscosity* are more significant than forces due to inertia.

lead A *control action* used to compensate for *lag* in a *process*.

LEL Lower explosive limit. A condition of an atmosphere in which any increase in temperature, fuel concentration, or oxygen concentration will cause an explosion. Potentially *hazardous areas* are often monitored for LEL to detect a rise in hydrocarbon vapors (fuel) so as to de-energize electrical circuits and prevent sparks from causing an explosion.

linear differential transformer A device designed to produce an electric output proportional to the displacement of a movable core within the primary coil.

linearization The modification of a system so that its outputs are approximately linear functions of its inputs.

load change Any change in the *process* that is imposed on it by changes in an external uncontrolled but related system.

LOS Lock-out–stop switch. An electrical switch designed to be locked in the off position to prevent operation of an electrical device. This is not necessarily a motor *disconnect switch* which is used to ensure safe maintenance. An LOS need not open the power conductors to a load; it typically only disables the control circuit.

measured variable The quantity or condition that is measured and controlled.

motor Any electromechanical device for converting electrical energy into mechanical energy.

motor starter A specially designed switch assembly used to connect and disconnect electrical *power* from *motors*. Motor starters may be either manually or electrically operated. If they also contain *current*-limiting devices (*fuses* or *circuit breakers*), then they are referred to as combination motor starters.

NEC National Electric Code. A set of legally required rules designed to prevent unsafe installations of electrical systems. The NEC is developed by the National Fire Protection Association (NFPA) and is updated every three years.

negative Electrical *charge* polarity associated with an excess of electrons (same as a deficiency of protons).

offset error A small *process error* which results from *proportional control action* whenever a load changes.

ohm The basic unit of electrical *resistance* or *reactance*. See *Ohm's law*.

Ohm's law Fundamental law of *electricity* which relates *electromotive force* to *current* and *resistance*. States that one *volt* of *EMF* will cause one *ampere* of *current* to flow through one *ohm* of *resistance*. Written as $E = I \times R$.

orifice plate A disc or plate-like member with a sharp-edged hole in it, used in a pipe to measure flow or reduce static pressure.

pacing The common term used to describe the *ratio control* type of *feedforward control* when applied to chemical injection.

perforated plate Flat plate with series of holes used to control fluid distribution.

phase The relative timing between two electrical signals in an *AC* system. A common term used to refer to one or more conductors carrying electrical power with the same timing relationship from a source to the load. With three-phase power each phase has a different timing relationship.

photovoltaic The *process* by which some materials convert incident light to an *electromotive force*.

piezometer An instrument for measuring fluid pressure or compressibility of materials.

pole An electrical term meaning an electrical conductor (wire) or a magnetic term for the end of a magnet. Electrical switches are often referred to by the number of poles they have which indicates how many separate *current* paths (wires) they can switch.

positive Electrical *charge* polarity associated with an excess of protons (same as a deficiency of electrons).

power The rate of flow of electrical energy. Measured in *volt-amperes* or *watts*. Equal to the product of *EMF* and *current*. Written as $P = E \times I$.

power factor The ratio of the amount of *true power* to the total *apparent power* in an *AC* system. Power factors less than one are caused by the effects of *inductance* or *capacitance* in an electrical circuit.

process Any single aspect of a system that has definable inputs, a reaction among them, and an observable output.

process error The difference between the *controlled variable* and the controller *set point*. Also known as *deviation*.

process lag A time delay in a process response to an adjustment caused by the inertia of the *process*. A contributing factor, along with *dead time*, in the time required for a *controlled variable* to respond to a change in value of the *final control element*.

propeller meter   An instrument for measuring the quantity of fluid flowing past a given point; the flowing stream turns a propeller-like device, and the number of revolutions are related directly to the volume of fluid passed.

proportional (action or control)   The *control mode* which varies the controller's output in proportion to the *controlled variable*.

proportional band   The gain of a *proportional action* controller. Defined as the percentage of the *controlled variable*'s change that will cause the output to change from minimum to maximum (100 percent).

rangeability   The ratio of the maximum flow rate to the minimum flow rate of a meter.

rate (action or control)   See *derivative action or control*.

ratio control   A *feedforward control* mode that causes the controller's output to always have a fixed ratio (e.g., 5 times, 1.5 times, 2.3 times, etc.) to the controller's input signal. See *pacing*.

reactance   The effect of *inductance* or *capacitance* which resists changes in *current* or voltage, respectively, in an *AC* system. It is measured in *ohms*.

reactive power   The amount of *power* used to "fill up" the energy storage in the reactive components (i.e., *inductance* and *capacitance*) of an *AC* system.

reluctance   A measure of the opposition presented to magnetic flux in a magnetic circuit, analogous to *resistance* in an electric circuit.

repeatability   The closeness of agreement among a number of consecutive measurements of the output for the same value of the input under the same operating conditions, approaching from the same direction, for full-range traverses.

reset (action or control)   See *integral action or control*.

resistance   The property of a material to impede the flow of *electricity* through it. Measured in *ohms*. See *Ohm's law*.

Reynolds number   A dimension criterion of the nature of flow in pipes. It is proportional to the ratio of dynamic forces to viscous forces; the product of diameter, velocity and density, divided by absolute *viscosity*.

rotor   The rotating part of an *AC* electrical *motor*, a *turbine* or *propeller meter*.

safety switch   Synonymous with *disconnect switch*.

semiconductor   Materials which can be easily forced to carry *electricity* only under certain circumstances are called semiconductors.

sensitivity   A measure of how small a change in the input of a device that can be detected and reacted to.

set point   The desired value of the *controlled variable*.

signal conditioning   Processing the form or mode of a signal so as to make it intelligible to or compatible with a given device, such as a data transmission line.

sonic flowmeter   A device for measuring flow rates across fluid streams by either Doppler-effect measurements or time-of-transit determination; in both types of flow measurement, displacement of the portion of the flowing stream carrying the sound waves is determined, and flow rate calculated from the effect on soundwave characteristics.

stator   The nonrotating or stationary part of an *AC* electrical *motor*.

straightening vanes   Horizontal vanes mounted on the inside of fluid conduits to reduce the *swirl* ahead of a flowmeter.

strain gauge   A device which uses the change of electrical resistance of a wire under strain to measure pressure.

swirl   Rotary motion of the flow superimposed on the forward motion.

synchronous motor An electric motor which is designed to run at *synchronous speed*.

synchronous speed The speed of rotation of the magnetic field in an *AC* electric *motor*. It is a function of the *frequency* of the applied *power* and the number of *poles* in the motor.

telemetry Transmitting the readings of instruments to a remote location by means of wires, radio waves, or other means.

torque The mechanical force of rotation. Commonly measured in foot-pounds (newton-meters in the metric system).

transceiver A device capable of both sending and receiving that provides the electrical interface to the physical medium.

transducer An electromechanical device which converts a physical quantity being measured (such as temperature or pressure) to a proportional, stable electrical output.

transit time In a sonic transit time flowmeter, the time duration for the sonic pulse to travel from the transmitter to the receiver.

transmission lag The *dead time* specifically associated with transmitting the measured value of variable between its measuring instrument and the receiving instrument, such as a controller or recorder.

true power *Power* available to do useful work. This is the total power in a *DC* system but only a portion of the power in an *AC* system.

tuning The act of adjusting the *proportional band*, *integral factor*, and/or *derivative factor* in an *automatic controller*.

turbine A machine for converting fluid flow into mechanical rotary motion, using energy sources such as flowing steam, water, or gas.

turbine meter A volumetric flow measuring device using the rotation of a *turbine*-type element to determine flow rate.

turbulent flow A flow regime characterized by random motion of the fluid particles in the transverse direction as well as motion in the axial direction. This occurs at high *Reynolds numbers* and is the type of flow most common in industrial fluid systems. Flow in which forces due to inertia are more significant than forces due to *viscosity* and adjacent fluid particles are more or less random in motion.

upset An unpredictable load change to a *process*.

U-tube manometer A device for measuring *gauge pressure* or *differential pressure* by means of a U-shaped transparent tube partly filled with a liquid, commonly water; a small pressure above or below atmospheric is measured by connecting one leg of the U to the pressurized space and observing the height of liquid while the other leg is open to the atmosphere; similarly, a small *differential pressure* is measured by connecting both legs to pressurized space. A manometer consisting of a U-shaped glass tube partly filled with a liquid of known specific gravity; when the legs of the manometer are connected to separate sources of pressure, the liquid rises in one leg and drops in the other; the difference between the levels is proportional to the difference in pressures—for example, high- and low-pressure regions across an orifice or *Venturi*.

vector sum The algebraic sum of two or more vector quantities. Vector quantities have both magnitude and direction.

vena contracta The smallest cross section of a fluid jet which issues from a freely discharging aperture or is formed within the body of a pipe owing to the presence of a constriction, such as an *orifice plate*.

Venturi A constriction in a pipe, tube, or flume consisting of a tapered inlet, a short straight constricted throat, and a gradually tapered outlet; fluid velocity is greater and pressure is lower in the throat area than in the main conduit upstream or downstream of the Venturi; used to measure flow rate.

Venturi flowmeter   Measures flow rate by determining the pressure drop through a *Venturi* constriction, and is followed by an expanding section for recovery of kinetic energy.

VFD   Variable frequency drive. A *motor* speed controller that uses adjustment of the applied power *frequency* to affect *motor* speed control. Used with *induction motors*.

viscosity   Measure of the internal friction of a fluid or its *resistance* to flow.

volt (V)   The basic unit of electrical force. See *Ohm's law*.

volt-amp (VA)   The basic unit of total *apparent power* in an *AC* system.

volt-amp-reactive (VAR)   The basic unit of total *reactive power* in an *AC* system.

vortex-shedding meter   A flowmeter in which fluid velocity is determined from the *frequency* at which vortices are generated by an obstruction in the flow.

watt   Basic unit of *true power* in an electrical system. Equal to one joule per second.

wild variable   Any input to a *feedforward controller* that is not impacted, even indirectly, by the output from the controller. An example would be the flow signal used for *pacing* of chemicals in a water plant.

Index

NOTE: An *f.* following a page number refers to a figure. A *t.* refers to a table.

Absolute pressure, 102
Actuators, 143, 144*f.*
Admittance probes, 107, 107*f.*
Air terminals, 36
Alternating current, 21
AM tone. *See* Amplitude modulation tone
American Wire Gauge, 22
Ampacity, 22
Amperage, 20
Amperes, 20
Amperometric chlorine residual analyzer, 115–116, 117*f.*
Amplitude modulation tone, 137–138
Analog telemetry, 133
    PDM (pulse duration modulation), 132, 133–136, 135*f.*
    pulse count, 132
    pulse frequency, 132, 136
    variable frequency, 136–137
Angle valves. *See* Globe valves
ARCNet, 193
Atmospheric pressure, 7
Aurora borealis, 19
Automatic control, 2, 161, 162
    compared with manual, 165–166
    feedback, 161–162, 162*f.*, 163–164
    feedback control of chlorine contact channel, 164, 164*f.*
    feedforward, 161–163, 162*f.*, 166–168
    feedforward control of chlorine contact channel, 163, 163*f.*
    feedforward vs. feedback, 164–165
    process error, 163
    process variables, 163
    set points, 163, 173–174
    wild variables, 162
Automatic reset control. *See* Integral control
Automation
    defined, 1
    reasons for, 2
Averaging Pitot flowmeters, 89–90, 90*f.*
    advantages and disadvantages, 91
    installation, 90–91
    maintenance, 91
AWG. *See* American Wire Gauge

Ball valves, 151–152
Bellows elements, 102, 103*f.*
Bias control, 167
Bourdon tubes, 102, 103*f.*
Bubblers, 106–107, 106*f.*
Butterfly valves, 151, 151*f.*

Calories, 20
Capacitance, 34–35
Capacitor motors, 35, 44–45
Capacitor run motors, 44
Capacitor start motors, 44
Capacitors, 35
Cavitation, 153
Centrifugal pumps, 156
Chemical conveyors, 157, 158*f.*
Chemical feeders, 157
    belt, 159, 160*f.*
    dry, 157–159, 159*f.*, 160*f.*
    gravimetric, 157, 159, 160*f.*
    liquid, 157, 158*f.*
    rotary paddle, 159, 159*f.*
    screw, 159, 160*f.*
    volumetric, 157, 159*f.*, 160*f.*
Chlorine. *See also* Residual chlorine monitoring
    feedback control of chlorine contact channel, 164, 164*f.*, 165, 166*f.*
    feedforward control of chlorine contact channel, 163, 163*f.*, 164, 165, 166*f.*
Chlorine membrane probes, 115, 116*f. See also* Residual chlorine monitoring
Circuit breakers, 31
    magnetic, 31
    thermal, 31
Circuits, 21
Circular recorders, 127, 127*f.*
Closed loop control. *See* Feedback control
CO_2 buffering systems, 116, 117*f.*
Communication network, 131, 132
Communication protocol, 131
Communications. *See* Digital communication systems
Computers. *See also* Digital control systems, Software
    analog data, 184
    central processing units, 182
    contact–closure devices, 184
    controllers as, 166
    digital data, 184
    in feedforward control, 168
    improvement of process control, 179
    input/output systems, 182–183
    main memory, 182
    mass memory, 182
    operator interface stations, 183
    peripherals, 183–184
    pulse data, 184–185
    video display units, 183
Conductors, 19, 22
    and resistance, 20
Cone valves, 152
Continuous duty motors, 49

Continuous process control. *See* Modulating
  process control
Continuous ultraviolet spectrophotometer, 119
Control, 1–2
Control diagrams, 52, 53*f.*–61*f.*, 63*f.*–64*f.*, 65*f.*
Control mode, 168
Control terminal units. *See* CTUs
Controllers, 124–126, 125*f.*
  computers as, 166
  distributed control units, 185
  programmable logic controllers, 185
  remote terminal units, 185
  smart field devices, 186
Converters, 122, 129
Coulombs, 19
CTs. *See* Current transformers
CTUs, 132, 133
Current, 19–20, 21
  alternating, 21
  direct, 21
  maximum current ratings, 22
Current transducers, 111
Current transformers, 23, 110–111, 111*f.*
Current-limiting devices, 22

DCUs. *See* Distributed control units
Density of water, 6
Diagrams. *See* Control diagrams, Ladder diagrams,
  One-line diagrams, Process and instrument
  diagrams
Diaphragm elements, 102, 103*f.*
Diaphragm pumps, 156
Differential areas, 17, 17*f.*
Differential gap, 169–170
Digital communication systems, 188–189
  communication media, 193–194
  connectivity, 193
  continuous polling, 192
  EIA standards, 189–190, 191*t.*
  error detection and correction, 193
  high-speed networks, 193
  IEEE standards, 191
  layers of communications, 189, 190*f.*
  links, 191
  local area networks, 189, 190, 190*f.*
  modems, 191
  multiconnected networks, 192, 192*f.*
  multidrop (bus) networks, 192, 192*f.*
  networks, 191–192, 192*f.*
  Reference Model for Open System Interconnection
    (ISO model), 189, 190*f.*
  remote communications, 191
  remote multiplexers, 188–189
  reports by exception, 192
  ring networks, 191, 192*f.*
  standards, 189–191
  star networks, 191, 192*f.*
  wide area networks, 189, 190, 190*f.*
Digital control systems, 179, 180–181
  components and tasks, 181, 181*f.*
  computer peripherals, 183–184
  computers, 182–183
  control room, 195–196
  data forms, 184–185
  distributed control, 194–195
  distributed control units, 185
  future trends, 196–197
  and interrupts, 180
  microcomputers, 194–195
  programmable logic controllers, 185
  remote sites, 196
  remote terminal units, 185
  SCADA systems, 181
  smart field devices, 186
  software, 186–188
Digital telemetry, 133, 134*f.*
  bidirectional, 132
  CTU (control terminal unit), 132, 133
  and RTUs, 132–133
  unidirectional, 132
Direct current, 21
Disconnect switches (disconnects), 23
Displacement pumps, 154, 155–156
Distributed control units, 185
Down conductors, 36–37
Down counters, 167

Eddy current drives, 155
EIA. *See* Electronic Industries Association
Elapsed-time timers, 166
Electric current to pneumatic (I/P)
  converters, 122, 129
Electrical potential, 20
Electrical systems, 2
  components, 2
Electricity, 18
  alternating current, 21
  amperage, 20
  amperes, 20
  backup systems, 27
  basic principles, 18–21
  branch circuits (branches), 24
  calories, 20
  capacitance, 34–35
  capacitive motors, 35
  capacitors, 35
  charge, 19–20
  chemically forcing out electrons, 19
  circuit breakers, 31
  circuits, 21
  conductors, 19, 22
  coulombs, 19
  current, 19–20, 21
  current transducers, 111
  current transformers, 23, 110–111, 111*f.*
  current-limiting devices, 22
  defined, 18
  direct current, 21

disconnect switches (disconnects), 23
distribution, 21–27
electrical force, 18
electrical potential, 20
electromagnetic interference (EMI), 35, 37
electromotive force (EMF), 20
electrons, 18–19
EMI signals, 38
encasement, 32–33
equipment grounding conductors, 31
fire hazards, 29
free electrons, 18–19
frequency, 21
fuses, 31
ground faults (grounding shorts), 31, 32
ground grids, 30
ground path, 31
grounding, 30–31
hermetically sealed contacts, 32
hertz, 21
horsepower, 20–21
inductance, 34
in-plant distribution, 24–27
insulators, 19
joules, 20
kilowatts, 20
lightning protection, 36–37
load centers, 24, 28f.
lower explosive limit, 29, 30, 33
magnetic circuit breakers, 31
magnetically forcing out electrons, 19
maximum current ratings, 22
motor control centers, 26
motor starters, 26
ohmage, 20
ohms, 20
Ohm's law, 20
one-line diagrams, 21–22
overcurrent systems, 31–32
overtemperature systems, 31–32
personnel hazards, 29
photovoltaic forcing out of electrons, 19
physical laws, 19–21
physically forcing out electrons, 18–19
power, 20–21, 23
power factor, 33–35
reactance, 33–34
resistance, 20
return path, 21
safety, 27–33
semiconductors, 19
sensors, 110–111, 111f.
single-line diagrams, 21–22
static electricity, 19
substations, 24, 25f.
surge protection, 37–38
surges, 35–36
switchgear, 24, 26f.
thermal circuit breakers, 31
thermally forcing out electrons, 19
three-phase power, 23
transformers, 23–24, 24f., 25f.
transients, 35
uninterruptible power supply, 27, 38
utility service connection, 22–23
voltage, 20, 22–23
voltage conversion, 23–24, 24f.
voltage levels, 23
voltage spikes, 35
watt transducers, 111
watts, 20
windings, 23
wire ampacity, 22
wire insulation, 22
wires, 22
Electromagnetic interference, 35, 37
Electromagnets, 42, 47
Electromotive force, 20
Electronic Industries Association, 189
Electronic instruments, 121–122
    features, 123t.
    power supplies, 123–124, 125f.
Electrons, 18–19
Elevation head, 10, 11f.
EMF. *See* Electromotive force
EMI. *See* Electromagnetic interference
EMI signals, 38
Encasement, 32–33
Energy, 9, 17
Equipment grounding conductors, 31
Equipment status monitoring, 112
Ethernet, 193
Event counters, 167

FCEs. *See* Final control elements
Feedback control, 161–162, 162f., 163–164, 168
    differential gap, 169–170
    and feedforward, 164–165
    gain, 171–172
    gap-action control, 169–171, 171f.
    integral mode, 174–176, 175f.
    integral rate, 174
    on–off mode, 168–171, 170f., 171f.
    proportional band, 171
    proportional mode, 171–174, 172f., 173f.
    proportional-plus-derivative mode, 177
    proportional-plus-integral mode, 176
    proportional-plus-integral-plus-derivative mode, 178
    rate mode, 176–178, 177f., 178f.
    repeats-per-minute, 174
    of reservoir water level, 168–169, 170f., 171f., 173f., 175f.
    and set points, 173–174
    timing graphs, 168, 169f., 170f.
Feedforward control, 161–163, 162f., 166
    bias control, 167
    by computers, 168
    event counters, 167

and feedback, 164–165
function modules, 167
pacing, 168
ratio control, 167–168
timers, 166–167
Fiber-optic sensors, 119
Final control elements, 143, 144f. *See also* Pumps, Valves
chemical conveyors, 157, 158f.
chemical feeders, 157–159, 158f., 159f., 160f.
Final elements, 143
FLA. *See* Full load amperage
Float-operated level sensors, 105–106, 105f.
Flow, 8
laminar, 9, 9f.
turbulent, 9, 9f.
Flow tubes, 70f., 74
Flow velocity, 8, 9f.
and cross-sectional area, 8, 9f.
Flowmeters, 67
averaging Pitot, 89–91, 90f.
coefficient of discharge, 68
differential pressure, 67–68, 105
fittings, 98
flow straighteners, 98, 99f.
flow tubes, 74
fluid-dynamic, 67, 68
flumes, 68, 94, 96–98, 97f.
head meters, 67–68
insert flow tubes, 74
installation precautions, 98
magnetic, 76–79, 77f., 78f.
modified Venturis, 74
open channel flow, 94–98
orifice plate, 74–76, 75f.
piping configurations, 98
propeller, 80–84, 81f.
sensor output, 99
signal conditioning, 99–100
signal enhancement, 100
sonic, 84–86, 84f.
turbine, 80–84, 81f., 82f.
variable area, 67, 68, 92–94, 92f.
velocity, 67, 68
Venturi, 69–74, 69f., 70f., 72f.–73f.
vortex, 86–87, 87f., 88f.
weirs, 68, 94, 95–96, 95f., 96f.
Fluid mechanics. *See* Hydraulics
Flumes, 68, 94, 96–97
advantages and disadvantages, 97
Palmer-Bowlus, 97–98
Parshall, 97, 97f.
Force, 16–17, 17f.
and differential areas, 17, 17f.
Four-wire transmitters, 124
Free electrons, 18
Frequency, 21
Frequency shift keying, 137, 138
Friction factor, 16

Friction head, 13, 13f.
FSK. *See* Frequency shift keying
Full load amperage, 45–46
Function modules, 128–129, 167
Fuses, 31

Gain, 171–172
Gap-action control, 169–171, 171f.
Gate valves, 152, 152f.
Gauge pressure, 102
Globe valves, 152, 152f.
Ground faults (grounding shorts), 31, 32
Ground grids, 30
Grounding, 30–31
ground path, 31
Grounding electrodes, 37

Handshake, 193
Head, 6
elevation, 10, 11f.
and energy, 9
friction, 13, 13f.
negative, 7
pressure, 13
total, 10, 11f., 13–14
velocity, 10–13, 12f., 15
Hermetically sealed contacts, 32
Hertz, 21
Horsepower, 20–21, 41
Hydraulic friction, 5
Hydraulic surge, 5
Hydraulics
density of water, 6
hydrokinetics, 8–16
hydrostatics, 6–8
incompressibility, 6
properties of liquids, 5
specific gravity, 6
viscosity, 6
Hydrodynamics, 16, 16f.
differential areas, 17, 17f.
energy, 17
force in hydraulic systems, 16–17, 17f.
work, 17
Hydrokinetics, 8
elevation head, 10, 11f.
energy, 9
flow velocity, 8, 9f.
flow velocity and cross-sectional area, 8, 9f.
friction head, 13, 13f.
hydrodynamics, 16–17, 16f.
laminar flow, 9, 9f.
liquid flow, 8
measurements, 9
pressure head, 13
quantity flowing in straight pipe, no friction, 14
quantity flowing in straight pipe, with friction, 14–16, 15f.
static pressure, 9, 9f.

total head, 10, 11*f.*, 13–14
turbulent flow, 9, 9*f.*
velocity head, 10–13, 12*f.*, 15
Hydrostatic pressure, 6–7, 7*f.*
  effect of container shape on, 7, 7*f.*
Hydrostatics, 6, 8
  atmospheric pressure, 7
  effect of container shape on pressure, 7, 7*f.*
  hydrostatic pressure, 6–7, 7*f.*
  vacuum, 7–8
Hydroviscous drives, 155

IEEE. *See* Institute of Electrical and Electronics Engineers
Incompressibility, 6
Indicators, 126, 126*f.*
Inductance, 34
Induction motors, 42–43, 43*f.*
Insert flow tubes, 74
Institute of Electrical and Electronics Engineers, 191
Instrument diagrams. *See* Process and instrument diagrams
Instrumentation, 2
  defined, 1
  familiarity with, 3
  operations used for, 2–3
  primary, 3, 121
  secondary, 3, 121–129
  specific tasks for each, 3
Insulators, 19
  and resistance, 20
Integral control, 174–176, 175*f.*
Integral rate, 174
International Standards Organization, 189
Ion monitoring systems, 119
I/P converters. *See* Electric current to pneumatic (I/P) converters, 122, 129
ISO. *See* International Standards Organization
ISO model. *See* Reference Model for Open System Interconnection (ISO model)

Joules, 20

Kilowatts, 20

Ladder diagrams, 56, 57*f.*
  symbols, 65*f.*
Laminar flow, 9, 9*f.*
LANs. *See* Local area networks
LEL. *See* Lower explosive limit
Level sensors, 105, 108
  admittance probes, 107, 107*f.*
  float-operated, 105–106, 105*f.*
  pneumatic bubblers, 106–107, 106*f.*
  stage recorders, 106, 106*f.*
  ultrasonic, 107–108, 108*f.*
  variable resistance, 107, 108*f.*
Light scatter turbidity monitoring, 112, 113*f.*
Lightning protection, 36

air terminals, 36
down conductors, 36–37
grounding electrodes, 37
Lightning rods, 36
Linear, variable, differential transformers, 102, 103*f.*
Liquid flow, 8
Liquids, properties of, 5
Live zero, 122
Local area networks, 189, 190, 190*f.*, 193
Lower explosive limit, 29, 30, 33
LVDTs. *See* Linear, variable, differential transformers

Magnetic flowmeters, 76–77, 77*f.*
  advantages and disadvantages, 80
  electrode cleaning, 79–80
  installation, 79
  maintenance, 79
  troubleshooting flowchart, 78*f.*, 79
Manual control, 1–2
Maximum current ratings, 22
MCCs. *See* Motor control centers
Mechanical systems, 2
Microprocessors, 122
Modulating process control, 143
Motor control centers, 26
Motors, 41
  across-the-line starting, 47
  automatic control, 59–62, 59*f.*, 60*f.*, 61*f.*
  bidirectional starting, 48
  branch circuits, 47
  combination starters, 47
  contactors, 47
  continuous duty, 49
  control diagrams, 52, 53*f.*–61*f.*, 63*f.*–64*f.*, 65*f.*
  control relay contacts, 55, 55*f.*
  control relays, 54–56, 55*f.*
  current-source controllers, 51
  defined, 41
  direct on-line (DOL) starting, 47
  disconnects, 49
  dual voltage ratings, 46
  eddy current clutches, 50
  electrically latched relays, 55
  electrically operated starters, 47
  and electromagnets, 42, 47
  feeder protection, 47
  float-operated level switches, 60–61, 60*f.*, 61*f.*
  flow sensing switches, 60
  full load amperage, 45–46
  full voltage starting, 47
  H symbol, 60
  hand operated starters, 46–47
  hand-off-auto (HOA) switch and circuit, 59–60, 59*f.*, 60*f.*
  and harmonics, 51–52
  heaters, 46
  HH symbol, 63

holding contacts, 55
horsepower, 41
hydroviscous drives, 50
induction, 42–43, 43f.
insulation type, 48
interlocks, 62–63, 64f.
L symbol, 60–61
ladder diagram symbols, 65f.
ladder diagrams, 56, 57f.
liquid clutches, 50
LL symbol, 63
local-remote switches, 58–59, 58f., 59f.
LOS switches, 49
maintained contact switch, 54, 54f.
man-auto switches, 62
momentary (spring-loaded) contact switch, 54, 54f.
momentary start button circuit, 54, 55f.
motor trips, 46
multispeed starting, 48
normally closed level switch, 60, 60f.
normally open level switch, 60, 60f.
overload protection, 46
overloads, 46
pressure sensing switches, 60
reduced voltage starting, 47–48
relays, 47
rotors, 41, 42
selector switches, 58–59, 58f.
service factor, 48–49
single-phase, 41, 44–45
slip, 43
squirrel-cage induction, 42, 43, 43f.
starter circuit with one switch, 53, 53f.
starter circuit with two switches, 53, 54f.
starter contactor coil, 52–53, 53f.
starter operation mechanisms, 46–47
starter sizes, 47
starters, 46–48
starting current, 45
starting techniques, 47–48
starting torque, 45
starting voltage, 45–46
stators, 41–42
status indicators, 57–58, 58f.
symbols, 65f.
synchronous, 44
synchronous speed, 42
temperature sensing switches, 60
three-phase, 41, 47
three-wire control circuit, 55, 56f.
three-wire control circuit with two control locations, 56, 57, 57f.
three-wire control using two level switches, 61, 61f.
three-wire control using two level switches with lock-out stop switch, 62, 63f.
three-wire control using two level switches with lock-out stop switch and low-level interlock switch, 63, 64f.
timed overload, 46

torque, 42
two-wire control, 53, 53f.
variable frequency controllers, 51–52
variable speed motor control systems, 50–52
variable torque transmission systems, 49–50
voltage-source controllers, 51
wound-rotor induction, 42, 43, 43f.
wound-rotor motor controls, 51

National Electric Code, 21, 29–30
    on motor branch cirucits, 47
    on motor disconnects, 49
National Electrical Manufacturers Association, 45
    insulation type standards, 48
    motor starter sizes, 47
    type A, B, C, D torque curves, 45
National Fire Protection Association, 21, 36
NEC. *See* National Electric Code
Negative head, 7
Negative pressure, 7
NEMA. *See* National Electrical Manufacturers Association
NFPA. *See* National Fire Protection Association
Nondisplacement pumps, 154, 156, 156f., 157f.
Northern lights, 19

Ohmage, 20
Ohms, 20
Ohm's law, 20
On–off control, 168–171, 170f., 171f.
One-line diagrams, 21–22
Open loop control. *See* Feedforward control
Operators, 1
Orifice plate flowmeters, 74–75, 75f.
    advantages and disadvantages, 76
    installation, 76
    maintenance, 76
    vena contracta, 75
Overcurrent systems, 31–32
Overtemperature systems, 31–32

P&IDs. *See* Process and instrument diagrams
Pacing, 168
Palmer-Bowlus flumes, 97–98
Parshall, Ralph L., 97
Parshall flumes, 97, 97f.
Particle counters, 117, 118f.
Pascal's law, 6, 7, 16
PD control. *See* Proportional-plus-derivative control
PDM telemetry, 132, 133–136, 135f.
pH sensing systems, 113, 114f.
    flow-through type, 113, 115f.
    immersion-type, 113, 115f.
PI control. *See* Proportional-plus-integral control
P/I converters. *See* Pneumatic pressure to electric current (P/I) converters
PID control. *See* Proportional-plus-integral-plus-derivative control

PID control functions. *See* Proportional, integral, and derivative (PID) control functions
Piston pumps, 155–156
Pitot tubes, 9, 10, 12, 14
PLCs. *See* Programmable logic controllers
Plug valves, 152, 152*f.*
Pneumatic bubblers, 106–107, 106*f.*
Pneumatic instruments, 121–122, 124
   air compressors, 122–123, 124*f.*
   features, 123*t.*
Pneumatic pressure to electric current (P/I) converters, 122, 129
Pneumatic systems, 2
Position transmitters, 112
Potential, 20
Power, 20–21, 23
Power factor, 33
   and capacitance, 34–35
   and capacitive motors, 35
   and capacitors, 35
   defined, 33
   formula, 34
   and inductance, 34
   and reactance, 33–34
Predictive control. *See* Feedforward control
Pressure. *See* Absolute pressure, Atmospheric pressure, Gauge pressure, Hydrostatic pressure, Negative pressure, Pressure head, Pressure instrumentation, Static pressure
Pressure head, 13
Pressure instrumentation, 102
   absolute pressure (psia), 102
   bellows elements, 102, 103*f.*
   bonded strain gauge sensors, 104
   Bourdon tubes, 102, 103*f.*
   diaphragm elements, 102, 103*f.*
   diaphragm seals, 102, 104*f.*
   gauge pressure (psig), 102
   linear, variable, differential transformers (LVDTs), 102, 103*f.*
   variable capacitance pressure cells, 102–103, 104*f.*
   variable reluctance sensors, 104
   vibrating wires, 104–105
Primary instrumentation, 3, 121
Process, defined, 161
Process and instrument diagrams, 199–206
   abbreviations, 199, 204*t.*
   function designations for relays, 199, 202*f.*
   letters, 199, 200*t.*
   loop description, 205–206, 205*f.*
   standard instrument line symbols, 199, 203*f.*
   symbols, 201*f.*
Process control, 143, 194. *See also* Automatic control, Digital control systems
   and computers, 179
   manual vs. automatic, 165–166
Process error, 163
Process variables, 163
Programmable logic controllers, 185

Propeller pumps, 156
Proportional band, 171
Proportional control, 171–174, 172*f.*, 173*f.*
Proportional speed floating control. *See* Integral control
Proportional, integral, and derivative (PID) control functions, 125–126
Proportional-plus-derivative control, 177
Proportional-plus-integral control, 176
Proportional-plus-integral-plus-derivative control, 178
psia. *See* Absolute pressure
psig. *See* Gauge pressure
Pulse duration telemetry. *See* PDM telemetry
Pulse frequency telemetry, 132, 136
Pumps, 143, 154
   centrifugal, 156
   components, 154
   controls, 154
   diaphragm, 156
   discharge pressure control, 156, 156*f.*, 157*f.*
   displacement, 154, 155–156
   drivers, 154
   eddy current drives, 155
   hydroviscous drives, 155
   nondisplacement, 154, 156, 156*f.*, 157*f.*
   piston, 155–156
   propeller, 156
   rotary displacement, 156
   speed control, 154–155
   turbine, 156
   variable ratio pulleys, 155
   variable speed couplings, 155
   variable speed drives, 155

Radio telemetry systems, 139–141, 140*f.*
Rate control, 176–178, 177*f.*, 178*f.*
Ratio control, 167–168
Raw water quality monitors, 119
Reactance, 33–34
Reactive control. *See* Feedback control
Recorders, 126, 127, 127*f.*
Reference Model for Open System Interconnection (ISO model), 189, 190*f.*
Remote multiplexers, 188–189
Remote terminal units, 131, 132–133, 185
   in digital telemetry, 132–133
Repeats-per-minute, 174
Residual chlorine monitoring
   amperometric chlorine residual analyzer, 115–116, 117*f.*
   chlorine membrane probes, 115, 116*f.*
   CO_2 buffering systems, 116, 117*f.*
Resistance, 20
Resistance temperature devices, 109, 109*f.*
Return path, 21
Rotary displacement pumps, 156
Rotors, 41, 42
RTDs. *See* Resistance temperature devices
RTUs. *See* Remote terminal units

SCADA systems, 131, 181
   and spread spectrum radio, 141
Secondary instrumentation, 3, 121
   controllers, 124–126, 125f.
   electric current to pneumatic (I/P) converters, 122, 129
   electronic instruments, 121–122, 123–124, 123t., 125f.
   enhanced function modules, 128–129
   four-wire transmitters, 124
   function modules, 128–129
   indicators, 126, 126f.
   integrators, 128
   live zero, 122
   and microprocessors, 122
   multiply/divide modules, 128
   pneumatic instruments, 121–123, 123t., 124, 124f.
   pneumatic pressure to electric current (P/I) converters, 122, 129
   proportional, integral, and derivative (PID) control functions, 125–126
   recorders, 126, 127, 127f.
   signal standardization, 121–122
   summation modules, 128
   totalizers, 127
   two-wire transmitters, 124
Semiconductors, 19
Set points, 163
   resetting, 173–174
Shaded pole motors, 45
Signal conditioners, 143, 144f.
   diverter valves, 144, 145f.
   electric positioners, 146–147, 147f.
   electric switching circuits, 144, 146f.
   hydraulic positioners, 145–146
   modulating, 145–147
   pneumatic positioners, 145–146, 147f.
   two-state, 144
Signal standardization, 121–122
Single-line diagrams, 21–22
Single-phase motors, 41
   capacitor, 44–45
   capacitor run, 44
   capacitor start, 44
   shaded pole, 45
   split phase, 44
Sleeve valves, 153
Slip, 43
Smart field devices, 186
Software, 186
   application software, 186–188
   operating systems, 186, 187f.
Sonic flowmeters, 84–85, 84f.
   advantages and disadvantages, 86
   installation, 85
   maintenance, 85
Specific gravity, 6
Specific ion monitoring systems, 119
Speed transmitters, 112
Split phase motors, 44

Squirrel-cage induction motors, 42, 43, 43f.
Stage recorders, 106, 106f.
Static electricity, 19
Static pressure, 9, 9f.
Stators, 41–42
Streaming current monitors, 118, 118f.
Strip chart recorders, 127, 127f.
Supervision, 1–2
Supervisory control and data acquisition systems. See SCADA systems
Surface scatter, 113, 114f.
Swing check valves, 2
Synchronous speed, 42

Tachometers, 112
Telemetry, 131, 132f.
   analog, 132, 133–137
   bidirectional, 132
   and cable TV, 141
   and coaxial cables, 138–139
   communication media and channels, 138–142
   communication network, 131, 132
   communication protocol, 131
   components, 131
   and copper wiring, 138–139
   CTU (control terminal unit), 132, 133
   digital, 132, 133
   and fiber optics, 139
   hybrid communication schemes, 141
   PDM (pulse duration modulation), 132, 133–136, 135f.
   pulse count, 132
   pulse frequency, 132, 136
   and radio systems, 139–141, 140f.
   RTUs (remote terminal units), 131, 132–133
   and satellite links, 141
   SCADA systems, 131
   and spread spectrum radio, 141
   and telephone lines, 139
   and tone multiplexing, 137
   and trunking systems, 141
   unidirectional, 132
   variable frequency, 136–137
Temperature sensors, 109–110
   resistance temperature devices (RTDs), 109, 109f.
   thermocouples, 109, 109f.
   thermowells, 109–110, 110f.
Thermocouples, 109, 109f.
Thermowells, 109–110, 110f.
Three-phase motors, 41, 47
Time-of-day timers, 166
Timers, 166–167
Timing graphs, 168, 169f., 170f.
Token-Ring, 193
Tone multiplexing, 137
Torque, 42
   curves, 45
   sensors, 112
   starting, 45
Total head, 10, 11f., 13–14

Totalizers, 127
Transducers, 101
    current, 111
    differential pressure, 69, 72*f.*–73*f.*, 105
    watt, 111
Transformers, 23–24, 24*f.*, 25*f.*
Turbidity monitoring, 112–113, 113*f.*
Turbine flowmeters, 80–82, 81*f.*
    advantages and disadvantages, 83–84
    installation, 83
    maintenance, 83
    troubleshooting chart, 82*f.*
Turbine pumps, 156
Turbulent flow, 9, 9*f.*
Two-state process control, 143
Two-wire transmitters, 124

Ultrasonic level sensors, 107–108, 108*f.*
Uninterruptible power supply, 27, 38
Up counters, 167
UPS. *See* Uninterruptible power supply

Vacuum, 7–8
Valves, 143, 153–154
    actuator selection and piping configurations, 149, 149*f.*
    actuators, 148–150
    AWWA standards, 150
    ball, 151–152
    bodies, 150
    butterfly, 151, 151*f.*
    and cavitation, 153
    components, 150
    cone, 152
    electric actuators, 149
    equal percentage, 150, 151*f.*
    failsafe, 153
    feedback, 153
    gate, 152, 152*f.*
    globe, 152, 152*f.*
    hydraulic actuators, 149–150
    inherent characteristics, 150
    installed characteristics, 150
    linear actuators, 148, 148*f.*
    linear percentage, 150, 151*f.*
    plug, 152, 152*f.*
    pneumatic actuators, 149–150
    quick opening, 150, 151*f.*
    rotary actuators, 148, 148*f.*
    seating mechanisms, 150
    selection, 150–153
    signal conditioners, 144–147, 145*f.*, 146*f.*, 147*f.*
    sizing, 150
    sleeve, 153
    staffs, 150
    stems, 150
Variable area flowmeters, 92–93, 92*f.*
    advantages and disadvantages, 93–94
    installation, 93
    maintenance, 93
Variable capacitance pressure cells, 102–103, 104*f.*
Variable frequency telemetry, 136–137
Variable ratio pulleys, 155
Variable resistance level sensors, 107, 108*f.*
Variable speed couplings, 155
Variable speed drives, 155
Velocity, 8
Velocity head, 10–13, 12*f.*, 15
Venturi flowmeters, 69, 69*f.*, 70*f.*
    accuracy, 70
    advantages and disadvantages, 73–74
    beta ratio, 69
    differential pressure transducer, 69, 72*f.*–73*f.*, 105
    head loss, 70
    installation, 71
    maintenance, 71
    troubleshooting guide, 71, 72*f.*–73*f.*
Vibration sensors, 112
Viscosity, 6
Volta, 20
Voltage, 20
    conversion, 23–24, 24*f.*
    levels, 23
    stepping down, 23
    at utility service connection, 22–23
Volume, 8
Vortex flowmeters, 86–87, 87*f.*
    advantages and disadvantages, 89
    installation, 87–89
    maintenance, 89
    troubleshooting guide, 88*f.*
VSDs. *See* Variable speed drives

WANs. *See* Wide area networks
Water distribution systems, 1
Watt transducers, 111
Watts, 20
Weirs, 68, 94, 95–96, 95*f.*
    advantages and disadvantages, 96
    rectangular, 95, 96*f.*
    trapezoidal (Cipolletti), 95, 95*f.*
    triangular (V-notch), 95, 96*f.*
Wide area networks, 189, 190, 190*f.*
Wild variables, 162
Wound-rotor induction motors, 42, 43, 43*f.*
    controls, 51

Y-pattern valves. *See* Globe valves

AWWA Manuals

M1, *Water Rates*, Fifth Edition, 2000, #30001PA

M2, *Instrumentation and Control*, Third Edition, 2001, #30002PA

M3, *Safety Practices for Water Utilities*, Fifth Edition, 1990, #30003PA

M4, *Water Fluoridation Principles and Practices*, Fourth Edition, 1995, #30004PA

M5, *Water Utility Management*, First Edition, 1980, #30005PA

M6, *Water Meters—Selection, Installation, Testing, and Maintenance*, Second Edition, 1999, #30006PA

M7, *Problem Organisms in Water: Identification and Treatment*, Second Edition, 1995, #30007PA

M9, *Concrete Pressure Pipe*, Second Edition, 1995, #30009PA

M11, *Steel Pipe—A Guide for Design and Installation*, Fourth Edition, 1989, #30011PA

M12, *Simplified Procedures for Water Examination*, Second Edition, 1997, #30012PA

M14, *Recommended Practice for Backflow Prevention and Cross-Connection Control*, Second Edition, 1990, #30014PA

M17, *Installation, Field Testing, and Maintenance of Fire Hydrants*, Third Edition, 1989, #30017PA

M19, *Emergency Planning for Water Utility Management*, Fourth Edition, 2001, #30019PA

M20, *Water Chlorination Principles and Practices*, First Edition, 1973, #30020PA

M21, *Groundwater*, Second Edition, 1989, #30021PA

M22, *Sizing Water Service Lines and Meters*, First Edition, 1975, #30022PA

M23, *PVC Pipe—Design and Installation*, First Edition, 1980, #30023PA

M24, *Dual Water Systems*, Second Edition, 1994, #30024PA

M25, *Flexible-Membrane Covers and Linings for Potable-Water Reservoirs*, Second Edition, 1996, #30025PA

M26, *Water Rates and Related Charges*, Second Edition, 1996, #30026PA

M27, *External Corrosion—Introduction to Chemistry and Control*, First Edition, 1987, #30027PA

M28, *Cleaning and Lining Water Mains*, First Edition, 1987, #30028PA

M29, *Water Utility Capital Financing*, Second Edition, 1998, #30029PA

M30, *Precoat Filtration*, Second Edition, 1995, #30030PA

M31, *Distribution System Requirements for Fire Protection*, Second Edition, 1992, #30031PA

M32, *Distribution Network Analysis for Water Utilities*, First Edition, 1989, #30032PA

M33, *Flowmeters in Water Supply*, Second Edition, 1997, #30033PA

M34, *Water Rate Structures and Pricing*, Second Edition, 1999, #30034PA

M35, *Revenue Requirements*, First Edition, 1990, #30035PA

M36, *Water Audits and Leak Detection*, Second Edition, 1999, #30036PA

M37, *Operational Control of Coagulation and Filtration Processes*, First Edition, 1992, #30037PA

M38, *Electrodialysis and Electrodialysis Reversal*, First Edition, 1995, #30038PA

M41, *Ductile-Iron Pipe and Fittings*, First Edition, 1996, #30041PA

M42, *Steel Water-Storage Tanks*, First Edition, 1998, #30042PA

M44, *Distribution Valves: Selection, Installation, Field Testing, and Maintenance*, First Edition, 1996, #30044PA

M45, *Fiberglass Pipe Design*, First Edition, 1996, #30045PA

M46, *Reverse Osmosis and Nanofiltration*, First Edition, 1999, #30046PA

M47, *Construction Contract Administration*, First Edition, 1996, #30047PA

M48, *Waterborne Pathogens*, First Edition, 1999, #30048PA

M50, *Water Resources Planning*, First Edition, 2001, #30050PA

M51, *Air-Release, Air/Vacuum, and Combination Air Valves*, First Edition, 2001, #30051PA

To order any of these manuals or other AWWA publications, call the Bookstore toll-free at 1-(800)-926-7337.

This page intentionally blank.